FROM THE ARAB OTHER TO THE ISRAELI SELF

Studies in Migration and Diaspora

Series Editor:
Anne J. Kershen, Queen Mary University of London, UK

Studies in Migration and Diaspora is a series designed to showcase the interdisciplinary and multidisciplinary nature of research in this important field. Volumes in the series cover local, national and global issues and engage with both historical and contemporary events. The books will appeal to scholars, students and all those engaged in the study of migration and diaspora. Amongst the topics covered are minority ethnic relations, transnational movements and the cultural, social and political implications of moving from 'over there', to 'over here'.

Also in the series:

London the Promised Land Revisited
The Changing Face of the London Migrant
Landscape in the Early 21st Century
Edited by Anne J. Kershen
ISBN 978-1-4724-4727-2

Migration Across Boundaries
Linking Research to Practice and Experience
Edited by Parvati Nair and Tendayi Bloom
ISBN 978-1-4724-4049-5

Human Exhibitions
Race, Gender and Sexuality in Ethnic Displays
Rikke Andreassen
ISBN 978-1-4724-2245-3

The Somatechnics of Whiteness and Race
Colonialism and Mestiza *Privilege*
Elaine Marie Carbonell Laforteza
ISBN 978-1-4724-5307-5

Antisemitism and Anti-Zionism
Representation, Cognition and Everyday Talk
Rusi Jaspal
ISBN 978-1-4094-5437-3

From the Arab Other to the Israeli Self

Palestinian Culture in the Making of Israeli National Identity

YONATAN MENDEL
Hebrew University of Jerusalem, Israel, and
University of Cambridge, UK

RONALD RANTA
Kingston University, UK

ASHGATE

Published by
Ashgate Publishing Limited
Wey Court East
Union Road
Farnham
Surrey, GU9 7PT
England

Ashgate Publishing Company
110 Cherry Street
Suite 3-1
Burlington, VT 05401-3818
USA

www.ashgate.com

British Library Cataloguing in Publication Data
A catalogue record for this book is available from the British Library.

The Library of Congress has cataloged the printed edition as follows:
Mendel, Yonatan, author.
 From the Arab other to the Israeli self : Palestinian culture in the making of Israeli national identity / By Yonatan Mendel and Ronald Ranta.
     pages cm -- (Studies in migration and diaspora)
 Includes bibliographical references and index.
 ISBN 978-1-4724-4935-1 (hardback) -- ISBN 978-1-4724-4936-8 (ebook) -- ISBN 978-1-4724-4937-5 (epub) 1. Palestinian Arabs--Israel--Ethnic identity. 2. Israel--Ethnic relations. 3. National characteristics, Israeli. 4. Cultural pluralism--Israel. 5. Palestinian Arabs--Israel--Social conditions. 6. Palestinian Arabs--Israel--Attitudes. I. Ranta, Ronald, author. II. Title.
 DS113.7.M46 2016
 305.892'74--dc23

                                                              2015029348

ISBN 9781472449351 (hbk)
ISBN 9781472449368 (ebk – PDF)
ISBN 9781472449375 (ebk – ePUB)

MIX
Paper from
responsible sources
FSC® C013985

Printed in the United Kingdom by Henry Ling Limited, at the Dorset Press, Dorchester, DT1 1HD

Contents

For Kathryn and Ella

List of Figures

Foreword

For as long as Israel has been a state its leaders and institutions have been obsessed over its character and identity. The need to justify the establishment of 'a Jewish State in Erets Yisrael' (the Land of Israel) – using the words of the state's founder and first Prime Minister David Ben-Gurion – included bringing together various immigrant groups who, despite claiming a common Jewish heritage, had little in common, provided a strong impetus for the creation of unifying national symbols as well as a common, *local* Israeli culture and identity. Therefore, from the rise of Zionism in Palestine, and during the years leading to and straight after the creation of the state, nationalist Jewish thought was intertwined with the creation of a unique Jewish national identity, and a unique localised or *sabra* culture. This process included, among others, the 'revival' and transformation of the Hebrew language into a national, modern, colloquial, old-new language, used by the people who 'returned' to their old-new homeland (following Herzl's *Altneuland*); the construction of unifying Jewish national symbols, from the blue and white flag with the star of David to the creation of El-Al as an Israeli 'blue and white' national airline; and the establishment of the Israeli Defence Forces (IDF), 'the melting pot of Israeli society', which became a focal point for the emerging state and society.

One example that came to mind when thinking about Israel's obsession over its character and identity was the annual 'What is Israeli in Your Eyes?' surveys. Every year, and increasingly around the time of Israel's Independence Day, local Israeli media conducts surveys in which they ask predominantly Jewish-Israelis, 'What is Israeli in your eyes?' This question of 'Israeliness' is very much part of the mainstream discourse in Israel. From the former populist journalist and current political leader Yair Lapid, whose writing and political campaign included the question '*ma Israeli be-'einekha*' (lit. 'What is Israeli in your eyes?'),[1] to art performances[2] and television initiatives,[3] all of which seem to satisfy a greater need to define the nation's identity and character. These surveys, in which the issue of 'Israeliness' is constantly reiterated, are repeatedly used by the media and the public as 'proof' for the existence of an imagined Jewish-Israeli identity and culture, which includes specific traditions, items and norms. These surveys provide a good indication of the nation's desires and phobias, wishes and concerns, limitations and boundaries.

Although the surveys are aimed ostensibly at defining or reiterating 'Israeli' or perhaps more specifically 'Jewish-Israeli' culture and identity, we quickly realised that many of the cultural, social and gastronomical items and norms that were labelled as 'Israeli' were in fact connected to the Arab world and culture, or at least to the Middle East. One example of this is the centrality of 'Israeli' food in the surveys. We were surprised to find that food such as falafel and hummus, which originated in Arab and Arab-Palestinian food cultures, and shawarma and burekas (börek), which represent a more regional (for example Turkish) food culture, were all given as examples of Israeli food and food culture. In other words, food items that are considered by Jewish-Israelis to be symbolic of their unique and local culture and identity were exactly those that, one hundred years ago, were commonly referred to and accepted as either part of the Arab and Arab-Palestinian food cultures or as part of the wider Middle Eastern or Ottoman-Turkish food cultures.

We came to notice that this phenomenon was not only present in or representative of Israeli food. Apparently, some of what Jewish-Israelis consider as uniquely and quintessentially 'Israeli' cultural items and symbols, by and large originate from an Arab culture and history, and frequently from the indigenous Arab-Palestinian culture specifically. These items and symbols were in fact imitated and adapted from the Arab culture generally, and the indigenous Arab-Palestinian culture more specifically, to become the most obvious symbols of Jewish-Israeli 'localness' and 'indigenousness'. What is striking is that the adoption of Arab and Arab-Palestinian items and symbols

---

1   The question of 'What is Israeli in your eyes?' was famously put forward by Yair Lapid when he was the leading journalist for *Yedioth Aharonoth*, Israel's most popular daily. Lapid later used this question in his successful political campaign running for the Knesset in January 2013, in which his party Yesh Atid (lit. 'There is a Future') became a dominant 'centre' party in Israel.

2   See, for example, the art exhibition in the Holon Design Museum in 2014, titled 'What Is Israeli in your Eyes?' In Tal Amit, 'What is Israeli in your Eyes? The Graduates of 2014', *Design Museum Holon*, 19 June 2014: http://www.dmh.org.il/heb/magazine/magazine.aspx?id=224&IssuesId=13 (accessed: 18 October 2014).

3   See, for example, the 2013 survey of its viewers by Israel's Channel 2, who were asked to send photos representing 'What is Israeli in your eyes?' See Einar Barzilai and Yoel Glazer, 'What Is Israeli in your Eyes? The Channel 2 viewers choose', *Mako*, 16 April 2013: http://www.mako.co.il/news-israel/local/Article-6dd5bdee7e31e31004.htm (accessed: 18 October 2014).

stands in contradistinction to the prevailing negative attitudes Jewish-Israelis generally exhibit towards Arabs and Arab-Palestinians. It also contrasts with the general context of the socio-political life in Israel – including the unresolved Arab-Israeli violent conflict; the tense communal relations and general animosity towards the Arab-Palestinian citizens of Israel; and Israeli Orientalism, all of which bolster the feeling of 'Western' Jewish-Israeli cultural and technological superiority over the backward Arab 'Orient', and can be seen, for example, in former Israeli Prime Minister Ehud Barak's saying that Israel in the Middle East is just like 'a villa in the jungle'.

And Barak's statement is obviously only the tip of the iceberg. A number of obvious examples and disturbing metaphors have been used by prominent Zionist thinkers and politicians over the past century that provide a glimpse of Zionism's and Israel's attitude towards the Arab and Arab-Palestinian 'Other'. Theodor Herzl, for example, who is considered among the founding fathers of Zionism, wrote that one of the justifications for the creation of the Jewish state in Palestine, as well as one of its main purposes, was to be 'a wall against Oriental barbarism'. Additionally, the Zionist Revisionist leader, Zeev Jabotinsky, argued that 'we [European Ashkenazi Jews],[4] thank God, have nothing to do with the East'. The statements made by these Zionist leaders, as well as many other similar examples, exemplify the superior Jewish Zionist attitude towards the region, and towards its indigenous population and culture.

Anti-Arab sentiments and representations are manifold in Jewish-Israeli society, and they can be clearly seen in cultural products as well. These have varied from the image of the Arab as either the enemy or the criminal in many Israeli movies and works of literature, to the Arab as evil and stupid, for example, in the famous Israeli children book series *Ḥasamba* (Cohen, 1985: 25). These representations also include the depiction of the Arab in Israeli school textbooks as a 'devilish enemy', 'dirty', 'primitive', 'agitated' and 'aggressive', and the Arabs more generally as the 'thugs' and the Jewish-Israelis as their 'victims' (Bar Tal and Teichman, 2005: 170; Peled-Elhanan, 2012: 225). The perception of Arab and specifically Arab-Palestinian inferiority is also part of Jewish-Israeli popular culture and colloquial language. Take for example the prevalent expression of ''Avoda 'Aravit', meaning 'Arab labour' in Hebrew, which is used to depict a shoddy, unfinished, and imperfect piece of work. Lastly, even the word 'Arab' or the phrase 'Ya 'Aravi' (lit. 'you Arab') have become derogatory terms used by Jewish-Israelis; they can often be used as expressions of poor sport's play among children ('You play like an Arab'); they can also be used as a way of criticising one's look, such as women with regard to heavy use of make-up or jewellery, or

---

4 Ashkenazi (plural Ashkenazim) is a term that refers to Jews of European descent.

men of Mizrahi[5] origin.[6] In other words, and despite the lack of academic literature on the topic, it seems that in Jewish-Israeli society the term 'like an Arab' is almost automatically derogatory, in that it typically stereotypes the Arab people and labels them with almost exclusively negative characteristics.[7]

It is from this contradiction, therefore, the hatred and contempt shown towards the Arab culture on the one hand, and the creation of a Jewish-Israeli culture which was based to an extent on Arab aspects and elements on the other hand, that our research emerged. Living on and off in Israel, having friends from both Jewish-Israeli and Arab-Palestinian societies, and while conducting research into different aspects of Israeli society, history and emerging culture, we both shared the same puzzled feeling regarding the Arab-anti-Arab-Israeli culture, and the attraction-repulsion relationship that Jewish-Israelis have with the Arab world in general and with the Arab-Palestinians in particular. This encouraged us to dig deeper into this dichotomous relationship and into the various layers of Israeli identity, focusing on Israeli symbols and symbol-making. By pushing forward a cultural archaeology of Jewish-Israeli society we strove to reconstruct a fuller story of this society's ambivalent and complex relationship with the Arab and Arab-Palestinian 'Other', and in particular its relationship with its Arab and Arab-Palestinian 'Self'.

This Israeli ambivalence towards the Arab world in general and towards Arab-Palestinians in particular is not a new academic topic and previous studies are addressed in this book. But perhaps a more telling example of how Arabic culture permeates into Israeli culture comes from a lyric by the late singer-songwriter Meir Ariel that hummed in our ears throughout the writing on so many levels. According to Ariel in his song *Shir Ke'ev* (lit. *Pain Song*) 'At the end of every

---

5 Mizrahi (plural Mizrahim) is a collective term that refers to Jews of Middle East and North African descent.

6 See, for example the critique of Roee Hasan, an Jewish-Israeli poet and publicist, on the use of the term 'you look like an Arab' against him in Israel. In: Roee Hasan, 'I look like an Arab', *Ynet*, 14 November 2014: http://www.mynet.co.il/articles/0,7340,L-4453409,00. html; see also, the analysis of Ella Shohat regarding the Israeli movie *The House on Shlush Street*, in which the Jewish-Ashkenazi figure ridicules the Mizrahi one for 'eating onion and olives, just like an Arab!', in Ella Shohat [in Hebrew] *Israeli Cinema: East/West and the Politics of Representation* (Raanana: The Open University of Israel, 2005), p. 173; see also the word 'Aravi' (the Hebrew word for 'an Arab') in the 'The Racial Slur Database', in: http://www.rsdb.org/race/arabs (accessed 31 January 2015); and Amalya Blum's (2012) analysis on how the word 'Arab' in Hebrew has been charged with negative load. She argues, however, that in order to reclaim it, Israelis should use it in a positive way: Amalya Blum [in Hebrew], 'You can say Arab', *Haaretz*, 29 March 2012: http://www.haaretz.co.il/ opinions/1.1674645 (accessed: 31 January 2015).

7 As we will discuss in the book, in recent years the term 'Arab' has also been associated with images of authenticity and quality, though mostly with regard to food products, and only in the margins (such as the radical Mizrahi discourse in Israel and the attempts to strengthen the concept of Arab-Jews).

sentence you say in Hebrew, sits an Arab with a nargilah [shisha]'. This line, even though not fully recognised by Jewish-Israeli people, is perhaps a lyrical 'un-researched' representation of what we unearthed in this study by looking at food, language, state symbols, daily symbols, and popular and national culture.

It is therefore important for us at this early stage to clearly explain how we differ from other works – and first and foremost academic works – that dealt with this subject. While acknowledging that the idea of Zionist and Jewish-Israeli fascination with and appropriation of Arab and Arab-Palestinian cultures and traditions, and later on denial and rejection of this appropriation, has been discussed and mentioned by several writers,[8] we are adding a different layer to the existing debate. While most of the focus hitherto has been placed on the early Zionist period, especially the first and second *'aliya*s (Jewish immigration waves); on specific movements, for example *Ha-Shomer* (the Watchman) and the *Palmach* (the 'Strike Force' brigades of the Haganah Jewish-Zionist paramilitary force); on the images of the Bedouin (semi-nomadic Arab ethnic group associated with dessert dwelling) and the Arab *Fallah*s (farmers/peasants); or on specific items and practices, we wish to move beyond these and provide an overarching inter-cultural analysis of this phenomenon in Israeli society. Though we agree that the first and second *'aliya*s were formative periods in the formation and construction of Jewish-Israeli identity and culture, we see this phenomenon as an ongoing one and not one that is limited to a particular period. We believe that it has changed over time, an issue we highlight in our research, yet we argue that the framework of adoption through erasure has been a central element in the creation of Jewish-Israeli identity and national culture. Additionally, while accepting the centrality of particular movements, images, items and practices, we do not view this phenomenon as narrowly associated only with them. We accept that these movements and images were of great importance, for example the modern image of the Jewish-Israeli sabra is directly related to movements such as the Palmach and some romantic notions regarding the Bedouin, but these do not tell the whole story we present in this research. The ambivalent and complex relationship we discuss is not limited to a particular period, movement, or even to specific cultural items and practices, but is far more encompassing and of wider spread. All in all, by combining academic research with personal, daily and at times anecdotal example, we argue that it permeates into and is manifested in almost every aspect of Israeli life, even if acknowledged – but by and large kept unacknowledged – by Jewish-Israeli people. This complex and ambivalent relationship had and still has a direct influence on, among others, the Hebrew language, Jewish-Israeli identity, national symbols, food and culture, as we show in this book.

Personally speaking, focusing on food, language and national symbols was a natural decision for us: one of us is a former chef who worked in Michelin-starred

---

8   We will acknowledge and discuss these writers as well as their research, which we have made use of, in the relevant chapters.

restaurants while the other has still not given up on the dream of becoming one; one of us uses Arabic daily in his work and research while the other has still not given up on the dream of finishing his Arabic studies and mastering the language; and both of us have dedicated our time and efforts to the study of Jewish-Israeli society and politics. We particularly wanted to shed light on a new set of values and processes relating to Jewish-Israeli society, and to consciously avoid research that delves into the 'hard politics' of the state. We believe that it is exactly those things that we put into our mouth, the words that come of it, and the images they carry, which serve as the azimuth for our ambitious, perhaps even naïve, journey into the making of the most Jewish, and also the most Arab, genes of Israeliness.

Writing the book in 2015 – in light of the ongoing Israeli-Palestinian conflict, 67 years after 1948, and 48 years after 1967, following two bloody intifadas, and the three wars on Gaza, the last of which took place while we wrote this book – made it difficult and almost surreal to focus on the Arab 'other' that 'hides' in the Jewish-Israeli 'self'. Yet perhaps despite and due to this gloomy and unpromising context, we believed that this mission not only holds an academic importance and basis, but also an important socio-political one. We hoped, and still hope, that while going through the book's pages, and being exposed to the use of Arab and Arab-Palestinian structures and elements in the construction of a Jewish-Israeli culture and identity, one will become not only frustrated with the current Israeli blindness to this phenomenon, but also hopeful regarding the chances of making this a point of departure for forgiveness and reconciliation. In other words, helping the Jewish-Israeli come to terms with and re-encounter the 'Arab that sits with a nargilah', but also the many aspects of the Arab and Arab-Palestinian cultures it has adopted. In essence this would mean a recognition that hate of the 'other' implies, in this case, hate of oneself. We therefore wish that this book not only highlights the schism within Jewish-Israeli identity, but also provide a possible glimpse of hope in the midst of the Israeli-Palestinian ocean of despair.

Series Editor's Preface

Much has been written about national identity, and, as Lynda Colley has depicted in the case of Britain, its 'forging'. Whilst the evolution of a British national identity can be traced back to the beginning of the eighteenth century, other nations are much younger and, in some instances, their identities are still in the process of formation. One of the younger nations, one born in the aftermath of the Second World War, is Israel. In this book the authors deconstruct the accepted popular symbols of Israeli identity to demonstrate how they are rooted in a place and culture far older than the Israeli nation state; in other words in the Palestine of pre-1948.

From its earliest inception Israel has been a hybrid nation, accommodating as it has, descendants of late-nineteenth-century Zionists making *aliya*, those who escaped and/or survived the Holocaust and those that arrived, post-1948, to help build the fledgling state; their numbers swelled by more recent arrivals from North Africa and Russia. It is a nation peopled by immigrants from many continents. There is no doubt that the creation of a unifying identity to accommodate this melting pot has been a challenge. In this volume Yonatan Mendel and Ronald Ranta provide us with the cultural and historical background to symbols we have come to accept as uniquely symbolising Israel: its food, its music, its music, its fruits and flowers, and, of course, *evrit* – the modern Hebrew language. The authors' thesis is that many of these have Palestinian-Arab, rather than European-Jewish, roots.

For Israelis, their national symbols link the past with the present and people with nation. Mendel and Ranta show us how, what were originally 'borrowings' from the Palestinian Arab population of the pre-1948 period, gradually became adopted and eventually owned by Israel and Israelis. For example, Jaffa oranges were in production in Palestine long before the arrival of political Zionism, yet the fruit is now firmly associated with the state. The prickly pear, or to give it its Palestinian/Israeli name, *sabr(a)*, is now a national fruit. Jews born in what is now the State of Israel are called *sabra*, yet it was in the decade before the creation of their nation state, that Zionist settlers in Palestine began to use the designation, explaining that it epitomised Jewish pioneers, the pear being thick and prickly on the outside and soft and sweet on the inside. Indeed one of the reasons for young people migrating to Israel is their desire to have *sabra* children.

The illustrations in this book further enhance and reinforce our awareness of the diverse symbols which both Israelis, and those living beyond Israel's borders, recognise as distinctly part of that nation. Yet, particularly in the case of food, the authors' demonstrate how so many of these have their roots elsewhere. 'Hummus, falafel, Israeli chopped salad, pita, and kebab on a cinnamon stick', all are part of Middle Eastern food culture. Indeed, Mendel and Ranta go so far as to suggest that Israel does not have 'an identifiable national cuisine', but rather, perhaps not surprisingly, has one which is reflective of the melting pot nature of its population allied to the inevitability of the need to use local produce. In contrast, there is one powerful Israeli symbol which does not have its roots in the land or culture of Palestine, but rather in the religion of the Jewish nation state. The blue and white Israeli flag reflects the design of the Jewish prayer shawl, with a Star of David – a Jewish symbol dating from mediaeval times – in its centre.

This book operates on two levels. Firstly, it demonstrates the ways in which the culture and language of the ages-old region of Palestine have been used – and at times abused – in the creation of modern Israel. Fascinating little known facts about the food, music and flowers of the locality are by-products of the authors' voyage through the nation's symbols of identity. However, this is not all the volume offers. It also focuses the reader's conception of the entire process of nation forging. What are the building blocks upon which a country's identity is built, from where are they sourced and how readily are they acknowledged? And finally, how acceptable is the end product to both insiders and outsiders? This is a book which informs and stimulates on an under-researched subject. It is a definite bonus for all those engaged in migration studies.

Anne J. Kershen
Queen Mary University of London
Autumn 2015

Acknowledgements

This book is the culmination of several years of collaboration between the two of us that was borne out of our mutual interest in the study of Israeli society and the Arab-Israeli conflict, but not less important in our mutual passion for food. The combination between the two was the point of departure for our friendship and the journey that resulted in this book.

Yet the book would not have come about without the help and support we received from many colleagues, friends and family. They have all contributed to our research, some in long interviews and others with important anecdotes. We probably cannot thank all of those who have helped us along the way, but will do our best not to forget anyone. We would like to thank the researchers of culture and identity and the chefs that have helped us throughout the research. We would like to thank Chef Hussam Abbas, Chef Habib Daoud, Dr Tal Ben-Tzvi, Dr Zeina Ghandour, Dr Liora Gvion, Dr Dafna Hirsch, Chef Salaḥ Kurdi, Chef Aaron Livingstone, Prof. Moti Regev, Chef Daniel Soskolne, and many others, for their time and patience in providing us with their perspectives and thoughts.

We also owe a debt of gratitude to Sophi Elsamni, Dr Atsuko Ichijo, Dr Lior Libman, Prof. Yehuda Shenhav, and particularly to, soon to be Dr Quinn Coffey, Dr Kathryn Tomlinson, and Sophie Richmond for reading parts of our work and providing us with much needed criticism and constructive feedback on all fronts.

We would also like to acknowledge the support and help we received from Ashgate, and in particular Neil Jordan the commissioning editor. Neil followed this project from the beginning and was always quick to respond to our queries and to offer his support.

In addition, Yonatan would like to thank the Department of Middle Eastern Studies at the University of Cambridge and especially the Centre of Islamic Studies and its director Prof. Yasir Suleiman. Cambridge was the place in which Yonatan conducted his research into the securitisation of Arabic studies in the Israeli school system, and where he started to study processes of "de-Arabisation" among Jews. Yonatan also wishes to express his gratitude to the Franz Rosenzweig Minerva Research Center at the Hebrew University in Jerusalem and its Director, Prof. Yfaat Weiss, and the ISF (Israeli Science Foundation grant 633/12) for their support in his post-doctoral research into roots of Arabic studies among the Jewish community in British Mandate Palestine.

Note on Transliteration

In this book we use the system of Arabic and Hebrew transliteration as outlined in the guidelines of the *International Journal of Middle East Studies* (IJMES). Where names of people and places have standardised spelling in English, the IJMES system of diacritics has been dispensed with. When citing books and articles written in Arabic and Hebrew, the footnotes include the English translation, and we mention in brackets the original language from which we quoted.

The Arabic letters *hamza* and *ʿayn* are written throughout the book as ʾ and ʿ respectively. When transliterating names of people, and when applicable, we followed the way these people write their name in English. When such examples were not available, we followed the IJMES transliteration system.

All translations in the text are ours unless specifically stated otherwise. We believe that the particular style of language and choice of words used by the people we quote, or in the materials we found during our field work, is an important aspect of our analysis presented here. Therefore, when a translation from a Hebrew or an Arabic source is given in the text, we have included in brackets certain key words, using the appropriate transliteration, in order to respect and highlight the original term or word used.

Glossary

'Aliya: literally meaning 'ascent' is a religious and political term given to the immigration of Jews to Erets Yisrael/the Land of Israel. From the beginning of Zionism in 1882 and up until the establishment of the state of Israel there were a number of distinct Zionist immigration waves to Israel/Palestine.

Ashkenazi (plural *Ashkenazim*): deriving from the biblical figure of *Ashkenaz*, who became associated with the region of central Europe; in Biblical Hebrew, the word *Ashkenaz* meant 'Germanic-speaking areas'. The term has been in use since the early Middle Ages and has since become a generic name for Jews of European descent.

Bedouin: refers to nomadic and semi-nomadic Arab ethnic group that is associated with desert dwelling; from the Arabic word for desert (*bādiyah*). The Bedouin are associated, especially in the West, with bravery, tent dwelling and hospitality.

Fallāḥ: from Arabic, meaning farmer or 'agricultural labourer' living in the Middle East. Derives from the Arabic word *fallaḥa* meaning 'till the soil'.

Hayay: comes from Hebrew *hayayah*, meaning 'being'; broadly understood as way of life, way of being and social environment.

Mizraḥi (plural *Mizrahim*): a collective term that refers to Jews of Middle Eastern and North African descent. Derives from the Hebrew *Mizraḥ*, meaning East, and literally means Easterner.

Palmach: an elite combat unit established in the 1940s as part of the main Zionist paramilitary organisation the Haganah. The word is a Hebrew acronym for 'striking companies' (*Pelugot ha-Maḥats*).

Sephardi (plural *Sepharadim*): a term used originally to denote the descendants of the Jewish community which was expelled from Spain (in Hebrew *Spharad*) following the Reconquista of 1492 and who mostly settled in the Middle East and North Africa. In recent times the term has become synonymous for Oriental or Mizraḥi Jews.

Totseret Haarets: literally meaning produce of the land; Totseret Haarets was a pre-state Zionist policy of promoting Jewish only and boycotting non-Jewish produce.

Tsabar (*Sabra* in English): derives from the Arabic *ṣabār* meaning 'prickly pear', a plant that became symbolic of the landscape in Palestine. Within the Zionist framework, the term has been used to denote Jews born in Palestine/Israel before or after 1948, allegedly highlighting their 'localness'.

Introduction

Our journey started three years ago with a joint research project that looked at the role of food in the context of Israeli and Palestinian relations. The focus of that work was initially on the Arab and Arab-Palestinian contributions to Israeli culture and identity through the examination of a number of well-known food items, practices and traditions that had been adopted and later appropriated and celebrated as 'Israeli' by the Jewish-Israeli society (see, Mendel and Ranta, 2014). Our main finding was that Israeli food culture was shaped by its encounter with the Arab region's food culture, and more importantly, only following a process that we referred to as de-Arabisation, food items were entitled to be seen as truly 'Israeli'. That is to say, elements from the Arab and Arab-Palestinian food cultures were initially adopted and adapted by early Zionist settlers as desirable and useful, and mostly as *native* and *local*. These elements, through their appropriation and nationalisation, became synonymous with Jewish-Israeli food culture and national identity, while the Arab and Arab-Palestinian origins and contributions were either marginalised or totally erased.

Our food findings, as we were to learn, were only the 'tip of the iceberg'. Early on in the project it became very apparent to us that the phenomenon we were observing was not limited to food but was far more widespread, and included other areas, such as national symbols, culture and language. There was no single eureka or Newton's apple moment, but a large number of interesting findings, anecdotes, and stories that brought about this realisation. During our food research we realised, for example, that 'Jaffa Cakes', the quintessentially British biscuit-size cakes, with their weird layer of orange jelly were related to and named after the oranges and city of Jaffa. Yet despite Jaffa oranges being today one of Israel's national symbols, as well as one of its most famous production and export trademarks, there are no orange orchards in modern Israeli Jaffa. So where did this idea of Jaffa oranges come from? It was only when several of our sources and interviewees took us back to the pre-1948 Palestinian Jaffa, to times when oranges were actually growing on the trees there, that the full picture of how and when the world-famous 'Jaffa Cakes' were actually invented became clear. This also tied together the Palestinian Jaffa orange, the adoption and de-Arabisation of this orange variety, and its marketing and trademarking as Israeli through and through. This mixture, of an Israeli national 'symbol', of Arab-Palestinian roots, and of images that have 'changed' hands, became gradually the rule, not the exception, of the many national items, symbols and practices – including sandals, flowers and dance steps – that we studied. All of which went through the same process of being de-Arabised on their route to becoming the epitomes of Israeliness.

We became convinced that this phenomenon of cultural adaptation leading to appropriation also has wider implications, and therefore can be useful for researchers examining case studies further afield. The implications of this phenomenon, we argue, can be relevant and important in understanding relations between different groups in national struggles in the context of divided, post-colonial and settler-colonial societies. Following this, we believe this research has a crossover appeal to other post-colonial and settler-colonial nations, societies and cultures, in the Middle East and elsewhere. In particular our focus is on the formation and crystallisation of national identities and cultures that emerge out of the dialectical relationship between Zionist settlers, and later Jewish-Israelis, and Arab-Palestinians, which includes an examination of the important cultural and political hegemonic forces at play.

The realisation of the wider implications of our research – to other areas of Arab-Israeli and Israeli-Palestinian relations and case studies further afield – was an important impetus we had in mind when we started writing this book. But there were other reasons for writing this study as well. We have felt that for too long much of the research and writing on the Arab-Israeli conflict generally, and the Israeli-Palestinian conflict particularly, has been marred by a stagnant and binary view that categorises identities in a very narrow manner. This has in turn narrowed down the scope of research into these conflicts, as well as the political thinking regarding possible 'way out' formulas, which have remained stuck in rigid structures, such as one-state or two-state solutions. We also felt that the thinking regarding the complex web of Israeli-Palestinian and Arab-Israeli interactions, in the past, present and future, needs to be broadened to include more than simply lists of security and economic related items that would be studied and that could be highlighted as possible avenues for political breakthrough. We felt this was especially relevant as we conducted our research in Israel in a very puzzling situation: on the one hand, the context in which our research was taking place – the ongoing third war on Gaza and the peaking of general and political hatred towards Arabs that was taking place all around us.[1] On the other hand, and due to our research, we kept finding and 'bumping into' Israeli national and cultural symbols that were strongly related to the 'enemy' Israel was fighting. This cultural and positive proximity was in dissonance with the images that kept flowing out of the Hebrew-speaking television screen and the Hebrew-written newspaper pages. This, especially in light of the critical and destructive war period, motivated and encouraged us to continue and further our research into the rise and fall of Israel's Arab roots. We also wondered what kind of 'doors' could be opened if and when the Arab origins of Jewish-Israeliness were more openly discussed and acknowledged.

---

1   See, for example: Ben Sales, 'After Gaza Conflict, Israel's Arabs Fear Rising Discrimination', *Haaretz* 7 September 2014.

Another interesting and relevant question to pose, which we thought about as well, is to what extent is the reverse true? To what extent has Arab-Palestinian identity and culture, or more generally the identity and culture of occupied or indigenous people, been influenced by Israel/Zionism or the external occupier/ settler political power respectively? This is indeed a fascinating question as it complements the research we undertook in this book. For example, when looking into Palestinian food culture we indeed encountered many cases of cultural diffusion and influences from the Jewish-Israeli side. One of the more remarkable examples of cultural adaptation and appropriation was revealed to us by one of our interviewees, Salaḥ Kurdi, the chef of the Jaffan restaurant Jamila. As an Arab-Palestinian growing up in Jaffa, which is a Jewish-Arab 'mixed city',[2] Kurdi told us he was well familiar with hummus, though it was not a dish he frequently ate. According to him, his mother used to send him occasionally to get some hummus from local Arab-Palestinian artisan producers as a quick and cheap way of having a family lunch. As he grew older, Kurdi decided to enter the food industry and worked as a chef for several Jewish owned Mizrahi (Jewish oriental) restaurants that catered mostly to Jewish-Israeli customers. Most of the food served in these restaurants – and this is something we will indeed touch upon later in the book – despite being labelled as 'Mizrahi', was a mixture of different food cultures loosely based on Arab and Arab-Palestinian mezzes. In the Mizrahi restaurant scene, Kurdi made and served hummus for a mostly Jewish clientele. A few years ago Kurdi decided to open a modern Arab restaurant in Jaffa specialising in Arab-Jaffan food. In order to stress his departure from more traditional Arab-Palestinian food, or the Jewish-Israeli perception of Arab food, he decided not to serve hummus in any shape or form.[3] Yet closing the circle, Kurdi complained to us that now, 30 years after he first went to get hummus from the artisan Arab producers in his home city of Jaffa, and as he decided to open an Arab-Palestinian restaurant with no hummus, he is confronted head on by the journey that the hummus made from his childhood into his refrigerator: seeing his wife and kids eat the Jewish-Israeli mass produced, allegedly 'Arab', *khummus*.[4]

---

2   For further reading on the concept of 'mixed towns' in Israel, and on the way Arab-Palestinian cities have been almost totally emptied of their Arab population to become 'mixed', see: Dan Rabinowitz and Daniel Monterescu (2008) 'Reconfiguring the "Mixed Town": Urban Transformations of Ethnonational Relations in Palestine and Israel', *International Journal of Middle Eastern Studies* 40(2): 195–226; see also Haim Yacobi, *The Jewish-Arab City: Spatio-Politics in a Mixed Community* (Routledge, London, 2009).

3   Interestingly, at roughly the same time, the Israeli food industry, which had mass produced hummus from the early 1950s, started to label some of its hummus as Arab for marketing purposes, a point we discuss in our food chapter.

4   We differentiate *hummus*, the Arab pronunciation and item, from *khummus*, which is the Israeli pronunciation for this food item and represents its Israelisation.

There are indeed many other such cases of Jewish-Israeli influence on Arab-Palestinian food culture, for example the consumption of chicken or turkey Schnitzel (breaded escalope) and milk chocolate drinks by Arab-Palestinians in Israel (Gvion, 2012). This, obviously, is a phenomenon found also in other fields. In the realm of language, for example, and as highlighted in Mar'i's research (2013) titled *Walla Bseder: A Linguistic Profile of the Israeli Arabs*, one can realise the way Hebrew words and linguistic structures entered into and how they influenced the Arabic spoken by the Arab-Palestinian citizens of Israel. However, and despite the interesting 'mirror image' that these examples provide, we thought that arguing that the colonial power influences the indigenous population is not new and is definitely not controversial. The way in which the coloniser influences the colonised has been central to the work of several post-colonialist writers, for example Frantz Fanon's *Black Skin, White Masks* (1991) and Homi Bhabha's *The Location of Culture* (1997). This relationship and influence is in fact part of the cultural hegemonic structures and power dynamics that prevail in settler-colonial societies. In that regard, Richard Wilk (2006) argues that dominant cultures can choose the foreign elements they wish to adapt and appropriate. In contrast, colonised or dominated cultures, however, rarely have the means to resist the encroachment of cultural materialism and must deal with the new ideas, images and structures that are imposed upon them.

This has been true to a certain extent for the Arab-Palestinian citizens of Israel, especially since 1948, and for Arab-Palestinians in the West Bank and Gaza Strip since 1967. In these situations, Arab-Palestinians have been a captive market for Israeli manufacturers and a captive audience for Israeli cultural and linguistic products. This is in line with many other historical cases in which economic and political colonisation, resulting in economic dependency, have resulted in greater political and cultural 'pressures' on indigenous communities.[5] Therefore, and while many interesting examples and areas for research come to mind – including the way Hebrew has been perceived by the Arab-Palestinian citizens of Israel as both the language of the occupier and the language of social mobility and economic success – these topics relating to Israeli influence on Arab-Palestinian culture and socio-political behaviour are beyond the scope of this book.

Another decision we made was related to methodology. We decided against presenting and designing our research on the Arab-Palestinian influence on Israeli culture as part of a quantitative analysis. Despite the possible advantages, we were more interested in the process and meaning of this phenomenon than in the percentages and numbers that this kind of research might generate. Our aim is not so much to quantify and measure, therefore, but to highlight the existing perceptions and structures regarding Jewish-Israeli and Arab-Palestinian relations. Following this, we use a slightly different methodological framework

---

5   See, for example, Walter Hixson's analysis with regard to settler-colonial and Native Americans relations, in Walter L. Hixson, *American Settler Colonialism: A History* (New York, Palgrave Macmillan, 2013).

for each chapter, with the overall purpose of demonstrating, comparing and exploring, through the four different case studies we chose, our main argument and theory regarding the role and place of Arab and Arab-Palestinian cultures in the creation of Jewish-Israeli identity and culture.

Lastly, this research acknowledges that many important 'Israeli' national symbols, items and cultural practices are not connected with the Arab and Arab-Palestinian cultures but originate either with European Jewry or other processes. Yet, we reserve our analysis for those elements that fall into our subject matter – the place of the Arab and the Arab-Palestinian within Jewish-Israeli national identity and culture – which, as argued in this book, are elements that make up a considerable part of Israeli symbols and practices, if not their majority.

National Identity and Culture

Our research focuses on the subject of national identity and culture, and the way they are conceived and constructed. Briefly explaining our approach to these terms and the relationship between them is important, and a good place to start would be to ask when and how do nations come about? We agree with Smith (2002) that nations do not appear out of a vacuum, and that the social groups that become nations indeed share, or perceive that they share some commonalities. We also follow Uzelac (2010), who argued that the formation of nations is a process which social groups go through. In this regard, we view the formation of nations and the construction of national identities as a multidimensional and continuous process. This process brings about a coalescing of ideas, people and traditions. These ideas and traditions are disseminated through a vernacular language by social entrepreneurs and movements, which help bring about a communal sense of shared history, values and identity: an imagined community (Anderson, 1983). The process is aided by the writing of a national history and narrative, and the creation of myths, traditions and symbols to represent the nation (Hobsbawm, 1983). To support, maintain and reinforce the national idea, the nation is reminded of its nationality through its banal representation (Billig, 1995). That is to say, the nation is continuously reproduced through a wide range of elements, from coins and monuments to texts and images.

The construction and reproduction of national identities is also not merely a top-down affair. Individuals negotiate with and engage with their national identities in their daily lives. Through their daily and mundane actions individuals create, add and transform meaning (Edensor, 2002). National identities are thus never fixed, they constantly evolve; though, the idea of the nation is routinised and institutionalised through top-down propagation and everyday acts. National identities are constructed in opposition to and are influenced by social and political events. The nation is thus more than merely an abstract idea propagated by elites, it is also 'a way of seeing, doing, talking and being' (Fox and Miller-Idriss, 2008: 540). As such, representations of national identities provide people, but also institutions and elites, with what Michael Skey (2011) terms 'ontological security': the sense of belonging to a nation as a source of stability and comfort in a fast-changing world.

Even though our main focus in this book is on the construction of Israeli national identity and on everyday life, we recognise that these are manifestation and reflections of Jewish-Israeli culture. As a result, what we examine and define as Israeli culture can be more broadly understood as the Jewish-Israeli way of life (known in Hebrew as *Hayay*). We are not particularly interested in 'high culture' or Culture with a capital C, but in the way in which constructions of Israeli identity are manifested in everyday life. This means focusing on practices and traditions that have emerged organically 'from below' as well as on cultural products that have been mass marketed and appeal to the Israeli mainstream. Our conception of culture is therefore principally the mundane expression of the nation's identity and its related experiences and symbols, which are based on the power relations and dynamics that exist within society. In a similar fashion to Hall (1996), we see culture as the site at which issues of power and identity are negotiated, expressed and explored. Nonetheless, and in agreement with Edensor (2002), and other writers on everyday nationalism, we also view culture as a site in which the everyday manifestations of nationalism are expressed and negotiated with.[6]

In this context, it is also important for us to discuss the creation of a Jewish-Israeli national identity and culture not only within the framework of a nationalism discourse but also within a post-colonial one. Zionism, though a national ideology, is also an example and expression of a settler-colonial movement. We accept that it is a unique manifestation of this phenomenon, in that the Zionist settlers immigrated to what they perceived to be their historic homeland, yet we perceive it also as a branch of a European-oriented and influenced movement. In this regard, we define settler colonialism as the movement of people from one place to another in order to settle and reproduce. With respect to nationalism and culture, settler-colonialists immigrate to the colony in order to establish their own country, society and nation (Verencini, 2010).

The construction of the settler-colonial nation is done mostly through the marginalisation, and at times forceful eradication of the indigenous population of the land, and the transformation of the settlers into natives, by adopting and appropriating indigenous knowledge and symbols. These two processes are integral to the construction of settler-colonial national identity and culture and the writing of the settler historical narrative. Tied in with Charles Tilly's (1985) influential formulation that 'states make wars and wars make states', the process of taking over the land of Palestine and marginalising and removing the native

---

6 We acknowledge the great variety and diversity present in Jewish-Israeli culture and are referring mostly to what is considered or understood as the Israeli way of life, rather than manifestation of class, gender, and other forms of culture/identity, even though it is clear that these too are present.

Arab-Palestinians, in particular the Zionist project and experience in the period leading to and during the 1948 war, are of vital importance in understanding the creation of the Israeli nation as well as its national identity and culture.

The Self and the Other

Within our exploration of national identities, the subject of the self and the other is of great importance. Identified by Ozkirimli (2005) as one of the five dimensions of national discourse, it was also highlighted by Said, who provided the starting point for our discussion by arguing that 'the development and maintenance of every culture require[s] the existence of another, different and competing alter ego' (1995: 332). In this respect, it has been a sociological as well as a political truism that the construction and maintenance of cultural and national identities requires an *other* to relate to; though not always, in most cases this *other* is also described and imagined as antagonistic. It is important to note that these constructed identities are never monolithic or rigid, but are constantly re-examined and reproduced and reified in relation to unfolding events and developments. In developing Said's point further, our research examines how the Jewish-Israeli self is continuously constructed through its Arab-Palestinian *other*. This helps us highlight what seems to us a fascinated yet tragic process of internalising the *other* through its marginalisation and elimination. This process can be viewed more metaphorically as the *othering* of Arab-Palestinians and Palestine in the mind of Jewish-Israeli, and their internalisation and *selfing* through Jewish-Israeli culture in reality. In many aspects this way of examining Israeli and Arab relations mirrors the process of mimicry described by Bhabha (1997). While Bhabha discusses the way in which the colonised adopts, internalises and adapts traits of the coloniser, our focus is on the inverse relationship between the two; how the coloniser adopts, internalises and adapts the traits of those he colonises.

The construction of the self and the other has been integral to the history and evolution of Zionism. The ideology of the Zionist movement was born as a response to the failure of the European emancipation process to culturally and socially integrate Jews as well as to the rising Anti-Semitism of the late-nineteenth century. These two developments put a spanner in the attempts to radically transform and modernise Jewish life in Europe, in particular in the Pale settlement (the area in Imperial Russia in which permanent Jewish settlement was allowed). In this environment of 'gentile' rejection on the one hand, and difficulties in achieving lasting cultural and social renewal, Zionism appeared to provide a genuine and unique solution to the, at times literally, existential problems faced by European Jews. However, Zionism – or the idea of creating a Jewish homeland in *Erets Yisraeli* ('the Land of Israel') – meant different things to different segments of the Jewish population, based on their countries of origin, education and religiosity, cultural differences, degree of assimilation, Mizrahi or

Ashkenazi affiliation, and social and political principles, dreams and beliefs. Additionally, Zionism also had to compete with other alternatives to the political and social problems faced by Jewish populations, mainly in Europe, which included among others, immigration to the US, socialist-oriented solutions, assimilation, Reform Judaism, strengthening of the Yiddish language and culture, and religious Ultra-Orthodoxy. Interestingly, one of the main effective mechanisms that was eventually 'chosen' and employed by Zionists in order to advance their cause, compete with the other alternatives and modernise Jewish life, was through the construction of a new Jewish identity, which included the political and cultural monopolisation of the Hebrew language.

The emphasis on the Hebrew language, and the support for its revitalisation, was a development that stood at the heart of the Zionist cause. Hebrew was a language that was able to tie together a Jewish past with the Jewish present and hopes of a Jewish future. This was needed both for nation-building on a communal basis, but also as a justification for a Jewish *return* to its 'forgotten' roots, which included the forgotten spoken language, the forgotten land – Palestine/Erets Yisraeli, and the forgotten local practices, traditions and customs. On top of this, the revitalisation of the Hebrew language encapsulated the revitalisation of the Jewish people. Just like the Zionist ideals of 'the conquest of the labour' (*kibbush ha-ʿavodah*) or 'the conquest of the land' (*kibbush ha-adamah*), the Zionist movement was also dominated by thinkers who advocated for 'the conquest of the language' (*kibbush ha-śafah*). Unlike the Yiddish and Ladino languages, which were associated with 'the diaspora' – Yiddish with European languages and communities and Ladino with Jews expelled from the Iberian Peninsula in 1492 – Hebrew was not 'diasporic' in nature, and its concerted revival was seen as an independent, brave, original and patriotic act.

The construction and development of a new Zionist Jewish identity also necessitated the presence of *others* to relate to. These ranged from the diaspora Jew – viewed as culturally stagnant, passive and backward – to the European Gentile, who was viewed as antagonistic, but whom Zionists wished to resemble culturally. On this last point, one only has to read the founding father of Political Zionism, Theodor Herzl's books *The Jewish State* and *Altneuland*, to comprehend just how much Zionism was inspired by European culture, social norms, politics and ideologies. The construction of the diasporic Jew, and even more specifically, the religious diaspora Jew from the Pale settlement as the *other*, and the complicated relationship with the European Gentile, ultimately started 'the great chain of orientalism' (Khazzoom, 2003). In this process, Zionist Jews orientalised and essentialised the Ostjuden Jews (Jews from the Pale settlement); followed by the orientalisation of the Mizrahi Jews, by among others Ostjuden, whom they encountered in Palestine/Israel; and, finally, in their process of becoming a 'Western' 'European' society, which participates in the Eurovision song contest and whose football teams compete in the European champions league, Jewish-Israelis – Ashkenazi but also Mizrahi – could only become truly 'Western' by orientalising the Arab-Palestinian people. This entailed the combined processes of romanticisation, imitation and disgust that one finds in Said's *Orientalism* and that echo throughout the pages of this book.

These last points lead us to a more expansive view of the *other*. As argued by Bar-On (1999), the construction of the new Zionist Jewish identity was not done only in relation to an *other*, but also in opposition to the other-self, or the *other* within the *self*. This way of understanding the history and evolution of the new Zionist Jewish identity contextualises it within a wider Jewish historical and social development. In other words, it is impossible to understand modern Jewish-Israeli identity and its historic evolution without examining the dialectical relations it had with what it considered the *other* within the *self*, namely the diaspora Jew. This type of 'diasporic Jew' (lit. *Yehudi Galuti*) at times took on the role of the antagonistic *other*, to be essentialised and orientalised, and whom the 'new Jew' (lit. *Ha-Yehudi ha-ḥadash*) was constructed in opposition to. This opposition is often characterised as the negation of the *galut* (diaspora), as studied among others by Raz-Krakotzkin (1993). At other times, the diaspora Jew was part of the new Jew: almost all of the leaders of the *'aliyas* (waves of Jewish immigration to Palestine) and the main Zionist ideologues were products of the diaspora. They were mostly raised in, what they would consider diasporic Jewish homes, brought up on the pillars of Jewish religion and tradition, and most of their families were probably still living, physically and emotionally, within the Jewish diaspora. Moreover, the renewed Zionist emphasis on the Bible and on the Hebrew language was a product of Jewish enlightenment in the diaspora. The Zionist leaders did not free themselves completely from the shackles of the diaspora. Instead the diaspora had a profound effect on their lives and they modified and incorporated it into their new identities.[7]

Following this, it is also interesting to analyse the way European Zionist Jews (the 'new yishuv') perceived the pre-Zionist Jewish community in Palestine (the 'old yishuv'), as a source of inspiration, even though limited to being external and exotic. Borrowing Derek Penslar's metaphor (2007: 97), the newcomer Zionist Jewish immigrants looked at the pre-Zionist Jewish people they met in Palestine as if they were 'Galapagos Jews', as if they were ancient turtles that had always been there, who were neither advanced nor sophisticated but who carried an important scientific purpose: they were there. They were there and could thus provide the proof that this is the land to which Jews belonged, including the new Zionist Jewish arrivals and Jews still living outside of Palestine. A proof to Zionist Jews that this is 'our place', a justification for the Jewish 'return' and a licence to 'redeem' the place from the Arabs, whose historical 'role' of preserving the land had ended, and from whom the well-preserved 'Jewish' place-names, food items and pronunciation could now be taken over.

---

7   Raz-Krakotzkin (1993), in his 'Exile within Sovereignty' (lit. 'Galut be-tokh Ribonut') analyses this process brilliantly, shedding light on the way the Zionist concept of 'negation of the Diaspora' has resulted in a narrow perception and imagination of Jewish culture.

There are other examples of the other-self within Zionism. The most important being that of the Mizrahi Jew, who upon immigrating to the 'western-oriented' Ashkenazi-European dominated Israel, was forced to make a choice: either to be an Arab (as if this was at all an option) or a Jew. Being both an Arab and a Jew was not an acceptable option any more – in particular outside the confines of their specific communities. Thus, those who wanted to succeed in Israeli terms were forced to become as Israeli and as Jewish as possible and, in parallel, to demonstrate they had shed their 'Arabness'. As the Israeli establishment adopted an increasingly Orientalist orientation, immigrants from Arab countries, places that were considered to be Israel's 'enemies', faced unique dilemmas and challenges to their cultural identity. Many, for example, were forced to prove their Jewishness through a process, which has been termed *religionisation*. According to Shenhav (2006: 114), who coined the term, 'If the Ashkenazi entrance ticket to Israeli-Jewish nationalism was through secularisation, the Mizrahi entrance ticket was by the strengthening of their Jewish religious beliefs'.[8] According to this theory, their religious identifier served as a 'safety-net' demonstrating that they would not slip back into 'Arabness', and that they would not threaten the Jewish self with the Arab 'other'. Furthermore, the desire to negate their 'Arabness' and to pose as little threat as possible was, according to Shohat (1988), also the reason why the term 'Arab-Jewish' fell out of favour and instead 'softer' expressions such as *Mizraḥim* (Orientals), *Sephardim* (descendants of the Jewish community which was expelled from Spain), and *'Edot ha-Mizraḥ* (the Oriental communities) were more commonly adopted.

In this regard, and in a similar manner, the attitude of Zionism towards the Arab-Palestinian indigenous population was one of acute ambivalence. As Raz-Krakotzkin argued (2005: 169):

> This is not to say that Zionist thought ignored the Arab entity, or rejected it simplistically. Zionists developed a myriad of approaches to dealing with the so-called Arab question, and produced a variety of images of "the Arab". Most of these, however, remained within the confines of orientalist discourse: the East (as represented by the Arab population) was either violent, irrational, and evil, or authentic and antiquated …. Both images had been established in various ways in Zionist-Hebrew culture from its earlier stages … the image of the Arab as the embodiment of the ancient Hebrew was a salient possible model for Zionist values and culture: the noble savage meets the biblical Israelite. These cultural manifestations demonstrate the complex and dialectical position of Zionism toward the Orient as a source of inspiration …

---

8   See also his fascinating analysis of the subject in the chapter entitled 'How did the Arab Jews Become Religious and Zionist?', in Yehouda Shenhav, *The Arab Jews: A Postcolonial Reading of Nationalism, Religion, and Ethnicity* (Stanford, CA: Stanford University Press, 2006), pp. 77–109.

So far, we have examined the construction of a Jewish-Israeli national identity and culture through its juxtaposition with the 'other' and the 'Other-self'. Yet in this book we propose to break the boundary between the other, the self, and the other-self, and to provide a more expansive way of understanding and looking at the relation between 'the Jew' and 'the Arab'. We suggest examining such relations not as binaries but as a scale where a clear overlap between the self and the other exists. In other words, viewing Jewish-Israeli identity as one that was not constructed and maintained simply in opposition to the Arab and Arab-Palestinian, but that also incorporated them.

In order to exemplify this, we would iterate that the history of the new Zionist-Jew begins with the diaspora and evolves with the arrival to Palestine. The Arab-Palestinian in general, and the Bedouin and the *fallah* in particular, provide points of reference for the construction of the new Zionist Jewish identity, in a similar fashion to the diaspora and the Gentile world. Identities and cultures that can or should be imitated on the one hand, but that should also be viewed as antagonistic and differential on the other. Zionist immigrants were thus able to orientalise and romanticise the native Arab-Palestinians as well as view them as vehicles for the preservation of biblical Erets Yisraeli and thus as representative of the biblical Jew, while at the same time also view them as the antagonistic other and as competitors for the land, in a similar fashion to other settler-colonial groups.

The Book's Chapters

The book is divided into four main chapters examining language, national symbols, food and culture. Our first chapter ('How do you say Arabic in Hebrew?'), which also serves as a historical introduction to Jewish-Arab relations, examines the relationship between the Arabic and Hebrew languages. From terms of endearment and swear words to names of items and geographic locations, the importance played by the Arabic language as spoken by Arab-Palestinians in the creation and development of modern Hebrew is evident. This phenomenon is in stark contrast to the perception and description of the Arabic language by Jewish-Israeli people today. Strikingly, Jewish-Israeli students find it more difficult to study Arabic, for example, than English, despite the fact that Hebrew and Arabic are both Semitic languages, with great similarity in terms of verb conjugation, grammar rule, vocabulary, etc. Instead, and due to the political conflict, Jewish-Israeli people see Arabic as intimidating and complicated. This perception does not allow the possibility of acknowledging the great proximity between the language of the self and the language of the 'Other', and also makes it harder for Jewish-Israelis to acknowledge the influence that Arabic had during the process of 'reviving' or 'revitalising' the Hebrew language in the beginning of the twentieth century.

In the chapter we examine questions relating to language policy and language planning through the case study of the revival/revitalisation of Hebrew in Israel. We therefore begin with the place of Hebrew in Israeli society and, more specifically, on the dialectical relationship, and tension, between Hebrew and Arabic. More specifically, we take account of the different phases in the relationship between the Hebrew and Arabic languages: from the moment Arabic was romanticised by the early Zionist immigrants to the country, through to the idea that the study of Arabic 'is part of Zionism' (S.D. Goitein), leading on to the 'redemption' of Hebrew through Arabic as seen in the writing of Eli ʿezer Ben Yehouda, and up to the inferior and securitised place of Arabic in modern Hebrew culture.

Language can be seen as one of the most important markers and signifiers of the nation-state. Following our discussion on Hebrew and Arabic, we move on in our second chapter ('What's in a Symbol?') to examine the wider relationship between Israeli and Arab symbols. National symbols come in various forms, from monuments and statues to images and food items. In this chapter, we look at the history and evolution of some of Israel's most iconic and well-known national symbols, including Jaffa oranges, prickly pears, and Israel's national flora and fauna. Through analysing these symbols and others, and their history and equivalents in the Arabic language and in Arab-Palestinian culture, the chapter unveils the hidden Arab-Palestinian side of Israeli national symbols, which, paradoxically, represent the Israeli nation and nationhood. It will be demonstrated that in becoming the natives, early Zionist settlers adopted and later appropriated and transformed Arab-Palestinian symbols into Jewish-Israeli ones, to the point that they are no longer imagined as anything but Jewish, which in nowadays Israel means both Israeli and non-Arab. Furthermore, the chapter argues that these symbols are not only part of the mythology of the Israeli state, but are also important elements in the construction and representation of the modern Jewish-Israeli national identity.

Our study of Israeli national identity also leads us to the important role played by food. When visiting Israel and experiencing its street food and food perceptions, restaurants and food markets, it becomes clear that one of the most important elements in defining Israel, to its own people as well as to non-Israelis, is what we define within the concept of *food culture*. Through multiple waves of immigration, Jewish-Israeli food culture has become a mixture of flavours and ingredients, but also influences. Many of these influences, for example those of the Jewish diaspora kitchens, are openly acknowledged, sometimes with a pinch of ridiculing, especially in the case of some Jewish Eastern-European dishes, for example, gefilte fish. Not so, as we argue in our third chapter ('Digesting the Nation'), is the important contribution made by the Arab-Palestinian food culture, which exists, but has since been appropriated as 'Israeli', and is rarely acknowledged as Palestinian, let alone celebrated as such.

Examining the history and evolution of Israeli food culture, reading through Israeli recipes, cookbooks, and food writing and discourses, and following up with interviews conducted with leading chefs and food researchers, writers and critics – Jews and Arabs alike – in Israel, the chapter demonstrates that, for political and ideological reasons, the contribution of Arab-Palestinians to Israeli food culture has been by and large marginalised and overlooked, and at times even denied. The chapter surveys the mechanisms that enabled this process and the ways that were found to justify it, for example the way Arab-Palestinian food has been redefined in many cases as either belonging to Jews of the Arab world (the Mizrahim) or as part of historic biblical Jewish traditions. The chapter demonstrates how these methods, imagined or otherwise, have by and large helped Jewish-Israelis to disregard the fact that many of the most prominent Jewish-Israeli dishes and ingredients have either been appropriated or are based on the Arab-Palestinian food culture.

The appropriation of Arab-Palestinian cultural elements is not limited to symbols, food and language. When one delves deeper, Arab-Palestinian elements appear in many aspects of Jewish-Israeli culture, including music, architecture, dance and dress. Based on field work as well as interviews with artists and academics, our fourth chapter ('The Creation of an Arab Non-Arab Culture') brings in a range of examples from Jewish-Israeli folk and popular culture to demonstrate the importance of Arab-Palestinian culture to Jewish-Israeli identity and culture. We argue that it is impossible to understand modern Jewish-Israeli national identity and culture without taking into account the dialectical relationship these have had with the Arab-Palestinian people.

All in all, this book aims to go to the heart of the Jewish-Israeli and Zionist discourse and practice – from the first words that created modern Hebrew, through the first steps of Jewish folk dances, and up to the food items and symbols that have come to define Jewish-Israeli identity and everyday life. From each of these points, we started to trace the connection between the Arab-Palestinian culture and the foundations of the Jewish-Israeli national culture and identity. We focused on symbols and culture for a reason. We believe that the culture, language, national symbols and foods that are associated with the nation, are not meaningless or insignificant, but actually serve as 'black boxes' that can help us trace the journey different groups made into becoming a 'nation'. We do not think that these symbols and practices made their way into national prominence by coincidence. Therefore, by trying to find the uniting elements of Israeli identity and culture, we hope to shed light on the Israeli national being, its founding pillars and the skeletons that are buried underneath. This, we believe, will provide a modest contribution to research on the creation of national identities and culture, but also bring to the fore an important discussion on the Arab roots of Israeli identity. Perhaps by uncovering this, as Mahmoud Darwish once said, we will highlight that the Israeli people need not to hide or be intimidated by the Arab within. As a matter of fact, acknowledging this could be one major step towards the end of Israel's fear of the Arab other.

Chapter 1

How Do You Say Arabic in Hebrew? Hebrew's Modern 'Revival' and the Place of Arabic

A few years ago, on a visit to an academic conference in Jordan, I found myself speaking to a Jordanian friend about football.[1] I learned that parallel to the European Champions League, there is also an Asian Champions League, and that the Jordanian champions, Al-Faisali, were to play on the day of the conference against Al-Majd, a top Syrian club from Damascus. Two hours later, I was sitting with another friend in the Amman international stadium, cheering for the local Jordanian side surrounded by about 20,000 excited Arab football fans. As an enthusiastic Jewish-Israeli football fan, it was a unique experience for me on many levels. First, which is more to do with the culture of football and the related research on this culture, I felt how the game really connected people and brought them together. It was one of the greatest moments I have experienced in Jordan, and I felt much closer to these football fans – as their mood shifted from happiness to anger throughout the game – than I felt towards my academic colleagues at the conference. Second, it was this cultural or ethnographic experience in the Middle East, which I took part in (quite indirectly, I know) that has clarified to me quite bluntly, the extent to which Israel is disconnected from the region. It was a mere demonstration to the ongoing cultural exchanges that exist in the region, from international football matches to academic conferences, and in which Israel as a state and Jewish-Israelis as people, hardly exist.

The third insight I had, though, was about language. As the game went on, and while I was rooting for the Jordanian side, I sometimes forgot I was actually in Jordan, in an Arab country in the Middle East, and not in Israel. It was because the flood of football-related curses that were pointed, mostly at the opposition side's players, but sometimes also at Al-Faisali's players, were similar in tone to those uttered in the football matches I attended in Israel, but strikingly also in their use of vocabulary. It then hit me: the same curses used by the Jewish-Israeli supporters were used by the Arab supporters in Jordan. It was a 'cursed' revelation. It started with '*Ibkī 'alek*' that was shouted by one of the Jordanian supporters after a poor tackle was made by one of the local player; '*Yā ibn sharmūṭa*' said another one on a cheeky handball by an Al-Majd defender; '*Yil 'an abūh*' said another when the Jordanian keeper conceded a soft

---

1   As experienced by Yonatan Mendel.

goal; '*Kus ukhto*' added his friend, followed by '*khara*', '*zbāleh*' and ' *'ars*'; '*Tfū 'alek*' said another, when the referee gave an unjust free kick to the Syrian side; '*Yikhrab beitak*' added the guy next to him; then, when Syrian side scored their second goal, a rain of curses, both creative and rude, poured from the stands, '*yā kalb ibn kalb*', said one, '*akhu manyūka*' shouted another as hulls of sunflower seeds flew out of his mouth. The Syrian side won the game that day, but I was left pondering less about the quality of the Jordanian strikers and defenders, and more about the remarkable language phenomena to which I was so frankly exposed. I knew that as the Hebrew language was revived it experienced a love-hate relationship with the Arabic language, and that modern Hebrew slang takes a lot of words from Arabic, yet I never had the chance to be in a linguistic 'laboratory such as the one at the international stadium'.[2]

The use of Arabic curses and swearwords in modern contemporary Hebrew is an illustrative example of the inter-connected relationship between the two languages and people. This chapter examines this relationship by giving an overview of the encounter between Hebrew and Arabic in Palestine, focusing mostly on the crucial period that began with the rise of Zionist immigration to Palestine (end of the nineteenth century) and up to contemporary Israeli society. We will also highlight the place and status of Arabic in Palestine, among Jews and Arabs alike, before the rise of Zionism, as well as specifically analyse the ambivalent attitudes that Jews in Palestine, later Israel, have had toward Arabic and Arabic studies. These attitudes have ranged from romantisation and a desire to speak the language; to utilitarian by seeing Arabic as a tool to 'return' to Hebrew; to a wish to speak Hebrew 'only' and to distance Hebrew from Arabic; and in the last contemporary stage, to a general disregard to the old ties between Jews and Arabic, and an even stronger denial of the imbedded connection and similar roots between two languages that used to be known as two Semitic sisters.

As we argue, the journey that Arabic made – from the *lingua franca* of a substantial number of Jews and the region's language to the language of the Other (Mendel, 2014) – has kept the Arabic curses more common and more welcome in Israeli society than any other reservoir of words, while other forms of Arabic that have influenced the Hebrew language, have been hidden. Therefore, the process in which Arabic curses made their way into Israeli common-speak can shed light on the changing relations in Palestine/Israel between the two people.

Language and Nationalism: On the Role of Language in Culture and National Identity

Acknowledging that language and language studies are much more than morphological or syntactic actions is of course a given in any research that

---

2   'Football experiences in Jordan, April 2009', as written in the personal diary of Yonatan Mendel, one of this book's authors.

involves language studies in the twenty-first century. The political significance of languages has been mentioned by various researchers over the past few decades. One can refer to Fishman's study from 1972 in which the socio-political load of languages is described. According to Fishman (quoted in, Suleiman, 2004: 14):

> Language is not merely a means of interpersonal communication and influences. It is not merely a carrier of content, whether latent or manifest. Language itself is content, a referent for loyalties and animosities, an indicator of social statuses and personal relationships, a marker of situations and topics as well as of the societal goals and the large-scale value-laden arenas of interaction that typify every speech community.

Following this quote, it is perhaps easier to delve into later scholarship on language policy, language ideologies and language studies in conflict situations that our specific case study – that is, the relations between Arabic and Hebrew in Palestine/Israel – is concerned with.[3] Firstly, it is perhaps necessary to start by clarifying a few of the basic theoretical concepts that enable us to look at relations between society and language, and that can assist us when trying to understand how Hebrew was 'revitalised' or 'revived'. The notion of *language policy*, and the ideologies behind it, has an evident importance to this matter. According to Spolsky and Shohamy, the word 'policy' is needed as it helps to better understand what makes one form of language more popular, or accepted than other. They argue that 'policy' in this context is used:

> to refer to an explicit statement, usually but not necessarily written in a formal document, about language use ... [and to] cases where the maker of the policy has some form of authority over the person expected to follow it. (1999: 32)

We make use of this when analysing institutions that have pushed forward and have helped to 'cement' a specific version of Hebrew that became the official, and accepted, language of the state of Israel. Another important assertion is regarding the term *language ideology*, which is perceived as 'something like language policy with the policy maker left out ... what people think should be done' (1999: 34). This notion, of the almost hegemonic nature of language

---

3 For further reading on the various connections between Arabic language and national identity in the Arab world, see the work of Yasir Suleiman. See: Yasir Suleiman, *Arabic in the Fray: Language Ideology and Cultural Politics* (Edinburgh: Edinburgh University Press, 2013); Yasir Suleiman, *Arabic, Self and Identity: A Study in Conflict and Displacement* (Oxford: Oxford University Press, 2011); Yasir Suleiman, *The Arabic Language and National Identity: A Study in Ideology* (Edinburgh: Edinburgh University Press, 2003); Yasir Suleiman, 'Nationalism and the Arabic Language: A Historical Overview', in: Yasir Suleiman (ed.) *Arabic Sociolinguistics: Issues and Perspectives* (Richmond: Curzon Press, 1994), pp. 3–24.

ideology, is important for our understanding of the way Arabic was perceived by
the revivalists of Hebrew and later by the Israeli state officials.

Another important work related to our understanding of the ideological
angles and considerations relevant to the use of languages is that of Blommaert.
According to him, in any research about language we should consider the
ideological formation and articulation of languages. In Blommaert's words,
'language does not lead a life of its own ... The story of languages is a story of
people who use them, manipulate them, manufacture them, name them' (1999:
425). It is that 'story' that we address in this chapter as we analyse the place of
Arabic in the revival and construction of the Hebrew language, and in particular
the connection between human behaviour (in our case assimilation, appropriation,
erasure), cultural products and language.

Besides referring to language policy and ideology though, it is also important
to highlight research that deals specifically with the use of language in situations
of war and conflict. Pavlenko's work in this regard is one example of the direct
implications that shifts in national identity and socio-political allegiances have
on the perception of languages as well as on language studies. She recalls, for
example, her first encounter with English language studies in the Soviet Union in
1975, during the Cold War, when her teacher told the class that,

> [T]oday is a very important day in your life – you are starting to study English.
> Your knowledge of this language will prove crucial when we are at war with the
> Imperialist Britain and United States and you will have to decode and translated
> intercepted messages. (2003: 313)

Pavlenko demonstrates how the political situation can help to either highlight
'similarities' between languages or to portray them as totally opposite to one
another. She concludes through two opposing case studies that different countries
have different perspective on what is to be done when a language becomes 'the
language of the enemy'. On the one hand, several US states, backed by the federal
department of education, forbade the teaching of German – which was the most
widely taught foreign language in the US – following World War I. Following
the war, speakers of German had to show their 'shift' to English only, and the
society – in response to the idea of a nation-state – shifted from a multi-lingual
society to an English-only one. On the other hand, the Soviet Union's policy
regarding English studies during the Cold War was based on a need to further
'foreignise' the language through its securitisation, and focused on teaching
English mainly for military purposes.[4]

---

4   This distinction of course relates also to the historical context of each case: post-
First World War, German ceased to be the language of a major European power (Germany)
and Empire (Austro-Hungarian), whereas the US remained a major power throughout the
Cold War. If Germany had remained a security concern for the US post-First World War,

Attitudes to language are generally linked to political processes of distinct periods. For example, within the Ottoman Empire or the Roman Empire, there was one official language of government; however, communities within the empire were not forced to obey the one-language one-nation language policy in their day-to-day lives. With the rise of nation-states in the eighteenth century and more clearly in the nineteenth century, the notion of one language for one nation-state became more prevalent and eventually the norm, leading in many cases to the restriction of other languages (see, for example, restrictions on the use of Catalan during the Franco period in Spain, or the restrictions placed on Indigenous people using their native languages in Canada and Australia). In the last 150 years, the emphasis has been on choosing *one* national language for the service of certain nationalistic goals, an equation, which is increasingly recognised as an ideological 'red herring' (Hornberger, 2003).

Shohamy's research is very relevant for our understanding of this modern situation: '[W]hile language is dynamic, personal, free and energetic, with no defined boundaries, there have always been those groups and individuals who want to control and manipulate it in order to promote political, social, economic and personal ideologies' (2006: 3). According to her, the idea that there is one way to speak the language, one way to say one word, one way to pronounce the language's sounds, are all related to acts of control from the state, and to the fact that the current political order is what we know as *the modern nation state*. She argues that, in the context of nation-states, 'language turned into a symbol of political and national identity and belonging' (2006: 27). Elsewhere Shohamy stresses that

> [W]ith the emergence of the nation-state, a new situation was created whereby questions were raised about the boundaries of groups and especially about membership ... It was then [and through the encounter with the "other"] that language became one of the main identifiers of membership, and of inclusion, and exclusion. (2006: 25–6)

Scollon summarises that 'it has been essential for the modern state to be perceived as having political boundaries that are isomorphic with language boundaries that are themselves isomorphic with cultural boundaries' (quoted in Shohamy, 2006: 26).

The idea that language needs to 'embody' the spirit of the state and reflect the 'purity' of the nation is related to the emergence of the nation-state and the idea of the language state (Makoni, 1998; Pennycook, 2004). This results in the nation-state's need to elevate one language as national and separate it from other languages. This view of closed, fixed, 'pure' perceptions regarding languages does not reflect the language reality and disregards the fluid relationships between languages of people who co-habit in one area, let alone in situation of an ongoing

---

the US establishment would probably have promoted its securitisation as well as its use and teaching.

conflict (Shohamy, 2006). As argued by Weinrich (2008), the differentiation between languages and the disregard of the similarities between them are all related to political conflict between entities, and as a matter of fact 'a language is nothing but a dialect with a Navy'.[5]

With regard to our case study, it seems very valid to us that as soon as the two people stood in opposition to one another – as soon as the Jews were fighting the Arabs in Palestine, as soon as the two had different armies, the inter-linkages between Hebrew and Arabic were 'sentenced' to be between two distinctly different languages, and the connection between them was denied, overlooked or rejected. The Hebrew language was then seen as an essential and integral part of the new Jewish national identity and culture, while Arabic was designated the language of the 'other' used by enemy states and people.

Before Zionism: On Jews and Arabic

In order to fully grasp the process by which modern Hebrew became to be what it is today – a language that has shifted so much from the original biblical Hebrew that it should not even be called Hebrew but *Israelit* (lit. 'Israeli language')[6] – one has to acknowledge the power-relations in which it was conceived and the love-hate relationship it had with Arabic. To do this it is, therefore, important to gain some historical perspective on the pre-Zionist Jewish relationship with Arabic.

For many centuries, Jewish people enjoyed particularly close ties to the Arabic language. At times, Arabic was a *lingua franca* for the Jewish people who lived in the Middle East, a daily language for communication and trade, and a language of culture, intellectual debate and also religion. Indeed, for thousands of years, the centre of the Jewish world was located in various regions of the Middle East, from Jerusalem to Baghdad and Cairo. In pre-Islamic times, Jewish communities spoke the language of the ruling empire or regime of that time, be it Greek, Aramaic, Latin or Persian. Hebrew was also prevalent during these periods; it was, however, by and large, neither the first nor the dominant spoken language among the Jewish communities.[7] This gradually changed following the Arab invasions of the seventh

---

5   The original was in Yiddish: *a shprakh iz a dialekt mit an armey un flot.* For further reading see: Max Weinreich, *History of the Yiddish Language* (New York: YIVO Institute for Jewish Studies, 2008), p. 362.

6   For further reading, see: Ghil'ad Zuckerman [in Hebrew], *Israelit Safa Yafa (Israeli – A Beautiful Language)* (Tel Aviv: Am Oved, 2008).

7   There are though quite a few examples of 'medieval' Jewish compositions from Byzantium and the Middle East written in Hebrew, and there is a substantial corpus of medieval 'Eastern' correspondence in Hebrew preserved in the Genizah. See, for example: Ben Outhwaite, 'Lines of Communication: Medieval Hebrew Letters of the 11th century', in E.M. Wagner, B. Outhwaite and B. Beinhoff (eds), *Scribes as Agents of Language Change* (De Gruyter: Berlin, forthcoming).

and eighth centuries when Jewish communities increasingly came under the rule of different Arab-Islamic empires. As a result, and similar to the acquisition of Greek or Persian, Jews of the region underwent a gradual process of Arabicisation which included the adoption of the Arabic language and its accompanying culture.

These empires related to their Jewish subjects in a variety of different ways, some better than others. In all cases, though, Jewish and Christian minorities were under special *dhimmī* status and, as such, they did not enjoy the same rights as the Muslim majority. Nevertheless, while sectarian tensions did occur, Jewish communities enjoyed an official status, and using Lewis's words (1987: 3–4) 'a recognized status, albeit one of inferiority to the dominant group, which is established by law, recognized by tradition, and confirmed by popular assent, is not to be despised'. This established status, overall, served as an asset and was advantageous for Jews in Muslim lands.

Despite some of the restrictions placed on Jews, Jewish communities flourished during most of the Islamic period. Various scholars contend that from the ninth century onwards, 'the bulk of Jews lived and prospered among Arab Muslims, whether in Spain, North Africa, or the Middle East' (Maalouf, 2003: 30–31). This relatively positive experience was most famously exemplified in areas such as Abbasid Baghdad, Fatimid and Ayyubid Cairo, the city of Kairouan in Tunisia and Ummayad Spain (al-Andalus), where Jews enjoyed a 'gracious productive and satisfying way of life they were not, perhaps, to find anywhere else until the 19th century' (quoted from, Johnson, 1987: 177–8). This period was considered to be a unique period of harmony in relations between Jews and Muslims. Goitien characterises this period as the pinnacle of Jewish-Arab relations; he wrote that: '[N]ever has Judaism encountered such a close and fructuous symbiosis as that with the medieval civilisation of Arab Islam' (1955: 130). Lewis (1987: 77) felt similarly; he believed that the Jewish-Muslim linguistic and cultural symbiosis surrounding Arabic was 'not merely a Jewish culture in Arabic, but a Judeo-Arabic, or one might even say a Judeo-Islamic, culture'.

This period of stability and prosperity continued throughout the Middle Ages. This was especially true in the Iberian Peninsula, in Iraq, Iran and Central Asia, the Maghreb, Yemen and Bilād al-Shām (including Palestine, Lebanon and Syria). Jewish communities flourished and enjoyed a relatively better situation – socially and financially – than during the preceding period and in comparison with those living in Christian areas. Undoubtedly, Jewish communities enjoyed a reasonably egalitarian, and for the most part sustainable and productive relationship with their Muslim neighbours.[8]

---

8   For further reading about this period and the history of the Jewish people in the Middle East, see: Mark R. Cohen, *Under Crescent and Cross: The Jews in the Middle Ages* (Princeton: Princeton University Press, 1994); S.D. Goitein, *Jews and Arabs: Their Contacts through the Ages* (New York: Schocken Books, 1955); Daniel Frank (ed.) *The*

Spain (al-Andalus) is a particularly poignant example of the flourishing of Jewish life. Jewish life and society in Spain in the tenth and eleventh centuries thrived, with Jewish communities enjoying both economic well-being and cultural prosperity. This place and period in time was later characterised in Hebrew as '*Tor ha-Zahav*' or 'the Golden Age' of Jewish creation. In this significant era of Jewish thought and religious output, Arabic was the primary vehicle for the transmission of Jewish expression. Some of the most significant figures of Jewish thought lived during this period. They included the Jewish philosopher and physician Abū ʿImrān Mūsā ibn ʿUbayd Allāh ibn Maymūn al-Qurṭubī (also known as Maimonides, or in contemporary Hebrew *Ha-Rambam*: *Rabbi Moshe ben Maimon*); the renowned and prolific grammarian and translator Saʿīd ibn Yūsuf al-Fayyūmī (in Hebrew: *Rabbi Saʿadia Gaon*); poets like Abū Ayyūb Sulaymān ibn Yaḥyā ibn Jabīrūl (in Hebrew: *Shlomo ben Gvirol*); and scholars such as Abū Ibrāhīm Ismāʿīl ibn Yūsuf ibn Naghrīla (in Hebrew: *Shmuʾel ha-Nagid*); who were al primarily active in al-Andalus.

In a similar vein, out of the thousands of scholarly works created during this period by Jewish scholars, Arabic was, by far, the dominant language. Some of the texts were in Arabic using Arabic script, some were written in Arabic using Hebrew script, and some Jewish scholars wrote their studies in Hebrew using Arabic script. This dominance of the Arabic language not only influenced Jewish thought, but also Arab philosophy, and some of the more influential of these Jewish texts include *Dalālat al-ḥāʾirīn* ('The Guide for the Perplexed', Hebrew translation: *Moreh Ha-Nevokhim*); *Kitāb al-Ḥujja wa-al-dalīl fī naṣr al-dīn al-dhalīl* ('The Kuzari', Hebrew translation: *Sefer Ha-Kuzari*); Yanbūʿ al-ḥayāt ('The Source of Life', Hebrew translation: *Meḳor Ḥayim*); and *Kitāb al-durar* ('The Book of Pearls', Hebrew translation: *Sefer peniney ha-musarim ve-shivḥey ha-ḳehalim*). These texts have had a profound influence on Jewish thought over the centuries and, to this day, remain central to Jewish religious tradition.

Increasingly, Jewish literary output during this period was written and consumed in Judeo-Arabic (Arabic written with Hebrew letters, known in Hebrew as *ʿAravit-Yehudit*). This specific type of Arabic, common within the Jewish community, often included consonant dots from the Arabic alphabet that helped to accommodate phonemes that did not exist in the Hebrew alphabet.[9] Interestingly, Arabic served at the time as a prime Jewish language used by Jewish communities and, for example, when analysing religious sources used during the twelfth century, the bulk of Jewish religious writing and literature – almost 90 per cent

---

Jews of Medieval Islam: Community, Society, and Identity (Leiden: E.J. Brill, 1995); Bernard Lewis, *The Jews of Islam* (Princeton, NJ: Princeton University Press, 1987).

    9   For further reading, see: Zion Zohar, *Sephardic and Mizrahi Jewry* (New York: New York University Press, 2005); Joshua Blau, *The Emergence and Linguistic Background of Judaeo-Arabic: A Study of the Origins of Middle Arabic* (Jerusalem: Ben-Zvi Institute, 1981).

of the sources – were written Arabic.[10] On top of this, Arabic translations of the Bible were widespread within Jewish communities and by Rabbinic Jews. This even extended to Muslim communities; a significant work from this period was *al-Tafsīr*, Saʿīd al-Fayyūmī's translation of the Bible into Arabic (and in Arabic script) in the early tenth century (Griffith, 2013).

In addition to having a profound influence on Jewish thought and religion, Arabic was also the language of Jewish scientific works. Interestingly, Jewish scholars often used translations from Arabic in the development of scientific texts. Freudenthal (2011), for example, in his analysis of the intellectual preferences of Jewish scholars in Europe from the twelfth to the fifteenth century, found that the overwhelming majority of Jewish philosophical writers were consistently more interested in Arabic than in Latin when borrowing, translating and using existing scientific knowledge. Therefore, Arabic was pervasive in secular and religious topics alike among Jewish thinkers of this period.

The importance of this period for Jews and Arabic stems from the fact that the Jewish Arabic-speaking community that lived under Islamic rule in Spain, North Africa and the Middle East represented the largest, most active and most influential Jewish community in the world at that time. Arabic played a major role in this Jewish renaissance; it was the language of daily life and was the vehicle for Jewish expression of spiritual, literary and religious achievements. Lewis emphasises this point in his comparison of Jewish communities living under Muslim rule with those living under Christian rule in the Middle Ages. He notes that: 'The Jews who lived in Christian countries, that is in Europe, were a minority, and a relatively unimportant one … With few exceptions, whatever was creative and significant in Jewish life happened in Islamic lands' (1987: 67). The flourishing of Jewish life during this period of time is inseparable from the daily interaction with Muslim communities and the Arabic language.

In a striking contrast, the way in which modern Israel relates to Arabic today would have been neither possible nor fathomable during these earlier periods. While today there is a clear distinction between 'Arabic' and 'Jewish communities', between 'Arabic' and 'Jewish intellectual creation', and between 'Jewish communities' and 'Arab communities', all to do with the processes we describe in this book (on the cultural, symbolic and linguistic levels), back then, Arab-Jewish identity, and the hyphen between them, represented the completion of one identity, not a contrast between two binary oppositions. Arabic was associated with Jewish daily life along with literary, scientific and philosophical masterpieces.

---

10   Hebrew remained the dominant language in Jewish liturgical poetry. Nevertheless, Arabic was also used for this, as second language. It is also important to distinguish between the situation of Jews in Europe, and Jews in the Middle East. For Jews in Europe, Arabic texts were translated *en masse* into Hebrew from the middle of the twelfth century onwards, while Eastern Jews continue to read and write in Arabic. For further reading, see: Moritz Steinschneider, *Jewish Arabic Literature* (Piscataway, NJ: Gorgias Press, 2008).

The 'Golden Age', as it came to be known, ended in the fifteenth century with the *Reconquista* of the Iberian Peninsula. As noted by Goitein (1955: 10), this development represented the end of 'the most important period of creative Jewish-Arab symbiosis lasting about 800 years'.[11] Like their Muslim neighbours, Jews were required to convert to Catholicism, or forced to flee, and they began to be hunted by the Inquisition. In 1492, following the surrender of Granada, and according to the Edict of Expulsion, no Jews were permitted to remain within the Spanish kingdom. In an occurrence of major significance in Jewish history, most of the Jewish communities relocated to Islamic-ruled regions that were amicable to Jewish communities, such as those in the Middle East and North Africa. As *Spharad* means 'Spain' in Hebrew, this group of emigrants, and their offspring, would later be referred to as *Sephardim*, demonstrating the deep roots of these communities in Spain.

Some of these expelled Sephardic Jews immigrated to Palestine. There, they encountered a Jewish community which was only a few thousand in number[12] and which spoke local Arabic.[13] Initially, the transition did not go smoothly; the Sephardim and the local Palestinian-Jews represented two different 'types' of Jewish life. The Sephardim, who within a short period of time outnumbered and looked down on the locals, brought with them an air of superiority due to their Muslim-Spanish culture and its achievements. In what today can be viewed almost as an act of colonising arrogance, the Sephardim nicknamed the local Jews *Musta'ribūn* (in Arabic: those who became Arab). This rather polemical term was intended to denigrate Palestinian Jews and was one of the ways in which Sephardim emphasised their Spanish-oriented 'advantages', such as command of the Ladino language (a Jewish-Spanish religious dialect).[14] As the years went by, however, and especially due to the numerical dominance of Sephardim, the two communities reconciled and merged. This was due, in part, to the gradual adoption

---

11   This Arab-Jewish creation applies to the Jewish communities in Arabic-speaking Spain, but should not be generalised to include all Arab Jews.

12   We use this figure following the estimation of Moses ben Mordecai Bassola regarding the number of Jews in Palestine/Erets Yisraeli in the beginning of the sixteenth century. In: Yitzhak Ben Zvi, *Erets Yisrael and its Settlement during the Ottoman Era* (Jerusalem: Yad Yitzhak Ben Zvi, 1975), p. 150.

13   This was also the case from 1516, when the Ottoman era in the region began. Despite the fact that the official Imperial language was then Ottoman Turkish (which was written in Arabic script and contained significant vocabulary from Arabic as well as morphological structures) the regional *lingua franca* remained Arabic. See, for example, Cornelis Versteegh, *The Arabic Language* (Edinburgh: Edinburgh University Press, 2001), pp. 226–41.

14   For further reading on the Musta'ribūn, see: Jonathan Gribetz, 'Musta'ribūn', *Encyclopaedia of Jews in the Islamic World*. Executive Editor: Norman A. Stillman. Brill online, 2013.

by the *Musta'ribūn* of Sephardic customs in prayer.[15] Thus, the two groups created a joint Oriental Jewish community in Palestine: the Sephardim.

This community contrasted with another Jewish group in the country: the *Ashkenazim*. Originating in Central and Eastern Europe (*Ashkenaz* means 'Germanic-speaking areas' in Old Hebrew), Ashkenazim were considerably smaller in number than Sephardim during the early Ottoman period. The arrival of Ashkenazi Jews in Palestine (or *Erets Yisraeli* in Jewish terminology) was primarily motivated by religious belief. They gradually increased in numbers throughout the seventeenth to nineteenth centuries and tended to live in small Ashkenazi Jewish-only enclaves. By and large, they maintained their traditional religious and cultural way of life and characteristics, including the use of Yiddish (Jewish-German) language.

These two groups – the Sephardic (those who emigrated from Spain and also the local *'Musta'ribūn'*) and the Ashkenazi – together constituted the Jewish population of Ottoman Palestine, which Zionist historiography refers to as *Ha-Yishuv Ha-Yashan* (lit. 'The Old Settlement'). Their percentage of overall population in the region was marginal: about 2 per cent in 1800 (5,000 out of 250,000) and about 5 per cent in 1882 (25,000 out of 540,000).[16] Due to their distinct religious backgrounds and lifestyles, the Jewish communities were concentrated in four cities considered to be sacred to Judaism. Known in Hebrew as *Arba' 'arey ha-ḳodesh* (lit. the four sacred cities), these cities were Jerusalem, Safed, Tiberias and Hebron.

Despite their shared religious beliefs, the communities remained culturally and socially distinct and employed different means of supporting themselves. Socio-economically, the Sephardic community was quite independent and its members worked and lived in Jewish neighbourhoods as well as in mixed Jewish-Muslim areas. The Ashkenazi-European community, on the other hand, maintained a more conservative way of life, and was more dependent on donations from Jewish communities in Europe, who sent money in order to support their Jewish brothers and sisters in Erets Yisroel (Yiddish pronunciation).[17]

---

15   For further reading, see: Minna Rozen [in Hebrew], 'The Position of the Musta'rabs in the Inter-community Relationships in Erets Israel from the End of the 15th Century to the End of the 17th Century', *Cathedra* 17 (October 1980), pp. 73–101.

16   The exact percentage of Jews in Palestine prior to the rise of Zionism is unknown. However, it probably ranged from 2 to 5 per cent. According to Ottoman records, a total population of 462,465 resided in 1878 in what is today Israel/Palestine. Of this number, 403,795 (87 per cent) were Muslim, 43,659 (10 per cent) were Christian, and 15,011 (3 per cent) were Jewish. Figures taken from: Alan Dowty, *Israel Palestine* (Cambridge: Polity, 2008), p. 13. See also: Mark Tessler, *A History of the Israeli-Palestinian Conflict* (Bloomington: Indiana University Press, 1994), pp. 43 and 124.

17   These donations are known in Hebrew as *Kaspei ha-Ḥaluka*. See: Menachem Friedman [in Hebrew], *Society and Religion: The Non-Zionist Orthodoxy in the Land of Israel* (Jerusalem: Ben Zvi Institute, 1988), p. 7.

Consistent with this separation, the Sephardic and Ashkenazi communities also maintained separate school systems. Schools catering to the Sephardic community included, alongside Jewish studies, the study of professional and non-religious subjects. By the early-nineteenth century a number of schools of this nature were in operation. By the middle of the nineteenth century, literary Arabic (also known as Modern Standard Arabic) was taught at a considerable number of Sephardic community schools. As a result of their education system and the immigration of Arabic-speaking Jews to Palestine, knowledge of colloquial Arabic in this community was the norm and not unusual.[18]

The same was not true of Ashkenazi schools. Not only was Arabic not included in the educational system, it was actively rejected. The study of literary Arabic was viewed as a secular and everyday topic not fitting with the religious nature of these schools. An anecdote demonstrates the active rejection of Arabic: in 1879, Sir Moses Montefiore (a British Jewish banker and philanthropist who supported the Jewish community in Palestine) offered to pay 200 pounds sterling to advance the employment of an Arabic teacher in an Ashkenazi religious school. He believed that literacy in Arabic would be of benefit to students who did not manage to succeed in their religious studies but might be able to acquire a profession. His initiative, according to Katzburg (1966: 300), was soundly rejected by leaders of the orthodox Ashkenazi community, who even created a religious ban 'against the study of any external wisdom'. In comparison, the leader of the Sephardic community praised attempts to increase the study of literary Arabic, which by then, was popular with his community (Kinberg and Talmon, 1994; Snir, 2005, 2006).

Beyond the learning and teaching of Arabic, the language was also pervasive in the general public space in which Jews in Palestine lived. As highlighted by several researchers, Arabic was the *lingua franca* of the region, and proficiency in it – even on a basic colloquial level – was necessary for managing trade and professional relations outside the confines of the Jewish community. Therefore, by the end of the nineteenth century, a considerable number of Jews in Palestine – mostly Sephardim but increasingly Ashkenazi Jews as well – had proficiency levels ranging from basic communication to full fluency and mastery of the language.[19]

Despite their principled rejection of Arabic, however, the Ashkenazi community was eventually forced to recognise the utility of knowledge of Arabic. As they were – in the eighteenth and nineteenth centuries – about 1 per cent of the total population of Palestine, and were also smaller in size than the Sephardic community, it was almost impossible for them to maintain a completely isolated

---

18   See, for example: Naphtali Kinberg, and Rafael Talmon (1994) 'Learning of Arabic by Jews and the Use of Hebrew among Arabs in Israel', *Indian Journal of Applied Linguistics* 20(1–2): 37–54.

19   On the integration of the Sephardim into the local Arabic language and culture, see: Eliezer Ben-Rafael and Stephen Sharot, *Ethnicity, Religion, and Class in Israeli Society* (Cambridge: Cambridge University Press, 1991), p. 26; and Andrew Forbes and David Henley, *People of Palestine* (Chiang Mai: Cognoscenti Books, 2012).

existence. In other words, they were forced occasionally to communicate with the majority (Arabic-speaking) population. Therefore, and even though Yiddish remained their primary language, they began to acquire Arabic 'for intercommunity purposes' (Spolsky, 1999: 165).

This is somewhat surprising and runs counter to the widespread belief that the Ashkenazi community of Palestine was a segregated, Yiddish-speaking society, which hardly had any contact with their Muslim or Christian neighbours. But this situation was not frozen in time. As a matter of fact, the situation of Ashkenazi Jews during this period can be compared to that of Yiddish-speaking communities in contemporary Israel or the United States. While they live in Jewish Orthodox areas, members of the Yiddish-speaking community speak the majority's language – either Hebrew or English – at proficiency levels which range from basic to full mastery. Then as now, some knowledge of the dominant language is simply unavoidable.

Another reason for language integration among the Ashkenazi community is related to the special characteristics of Yiddish. The language was described by Max Weinreich (2008) as a 'fusion language'; it combines aspects of Germanic, Slavic, Semitic and other languages. The historically flexible nature of Yiddish facilitated the absorption of vocabulary and new linguistic structures from 'hosting' languages and societies. Even today this is the case, and in the United States, for example, Yiddish has transformed into *Yinglish* (Bluestein, 1989). Similarly, it is clear that contemporary Yiddish in Israel has been influenced by Hebrew (Korn, 1983).

The relationship between Yiddish and Arabic is described in Mordecai Kosover's (1966) research *Arabic Elements in Palestinian Yiddish*. The author highlights the gradual increase of Arabic words in Palestinian Yiddish, in the seventeenth, eighteenth and nineteenth centuries. According to Kosover, this process started in the early-seventeenth century, when Ashkenazi community leaders realised that lack of Arabic prevented them from participating in various trades. This, as Kosover shows (1966: 100), was highlighted in *Ways of Zion*, a book written by Moshe Porges in 1650, where the author writes: 'Some Sephardic Jews own shops and stores, full of all kinds for sale .... Only we, Ashkenazim, do not know the languages to converse with various people, and we are therefore unable to trade with them.' Kosover (1966: 99) therefore explains the penetration of Arabic elements into the Yiddish language in Palestine as instrumental and 'mainly *economic* in nature' and describes the economic incentive for proficiency in Arabic.

From the nineteenth century onwards, Kosover increasingly identified signs that the community acquired knowledge of Arabic, outside of the school system. Kosover cites a book, *Ḳoroth ha-'Ittim li[Ye]shurun b'Erec Yisrael*, written by Rabbi Menaḥem-Mendl of Kamenitz in 1839, following his immigration to Safed. One chapter is dedicated to 'Lashon 'Aravit' ('The Arabic Language'), where he mentions the crucial importance of Arabic. Menaḥem-Mendl concludes the chapter with probably one of the first Yiddish-Arabic dictionaries. He wrote:

Jerusalem is called in their language *'Ir Quds'* [*'Ir* means "city" in Hebrew], Nablus is *Nablāt*, Safed is *Sāfāt* … When one inquires about the price, he says *ķadesh hadā* … money is *mesāri*, and when one asks for money he says *hati mesāri* … *bācīl* [means] onions, *mīzan* means the scales, *zeit* is oil … and *ḥubzeh* is bread …. (1966: 106)

During the second half of the nineteenth century and at the dawn of the twentieth century, Arabic was becoming more common among Ashkenazi Jews than ever before. While still not taught at Ashkenazi schools, its penetration was related to socio-economic developments locally, and to developments relating to influences of the Enlightenment in Europe. Abraham Frumkin, who was born in Jerusalem in 1873, recalls in his memoirs the multi-lingual reality in which he grew up. He specifically mentions his command of Arabic writing 'I spoke [Arabic] very fluently while still a child. I acquired it without the slightest effort on my part. … The times were different then. No one yet knew of an Arab-Jewish problem …' (1966: 114). Similar sentiments were expressed by Ephraim Cohen-Reiss (born 1863). He remembers that 'the Ashkenazi Jews of Safed generally spoke Arabic far better than the Jews of Jerusalem. … Even in their way of living, as well as in the attire, they were closer to the Arabs' (1966: 114).

The close relationship between Arabic and Yiddish among Ashkenazi Jews in Palestine at the period under discussion, and the entry of Arabic words into the Yiddish spoken in Palestine, is demonstrated by Kosover in different parts of this book. Some examples from the world of food include the Yiddish sound *kuselakh*, which means courgettes (and comes from 'kūsā' in Arabic with the 'akh' Yiddish diminutive plural) and *kaftele* which means meatballs ('kufta' in Arabic). He also points to the Yiddish-Arabic word *koyes* (with its Yiddish and based on the colloquial Arabic term 'kweyyis' that means 'all right') and '*Allah Karim, Got vet helfn*' (God will help) in both Arabic and Yiddish. This symbiosis therefore took place on a number of levels.

All in all, this summary sets the socio-linguistic stage in Palestine on the eve of the rise of Zionism in 1882. It has introduced some of the processes, definitions and social groups addressed by the research and, we believe, highlights the depth of the change that occurred in the twentieth century. It is important to remember that, at the end of the nineteenth century Jews in Palestine were a small minority that lacked contemporary nationalistic sentiments, and did not have a standing army, a national anthem or a national language. The Jewish communities outlined here integrated Jewish religious practices with everyday life while also maintaining good relations with their neighbours and had a basic to high level proficiency in Arabic. Unlike today, violent incidents between Jews and Palestinians were few in number and dissimilarities between the communities were not a source of tension or an excuse for conflict. In Frumkin's words, 'No one yet knew of an Arab-Jewish problem'.

Between Three Conquests: Labour, Land and Language

The year 1882 was an important landmark in the history of Palestine, with the arrival of the first wave of Zionist immigrants from Europe. The national aspirations embedded in Zionism characterised this wave of immigrants and, more significantly, the Jewish immigration that followed. The dominant Zionist political line positioned the Jewish national interests in Palestine in opposition to those of the Arab-Palestinians, and signalled the beginning of a zero-sum process that eventually changed the demographic, social and political situation in Palestine. Its most significant result was the creation of Israel in 1948 and the expulsion of approximately 700,000 Palestinians (Morris, 2004; Pappé, 2006; Shlaim, 2001).

The Zionist immigration to Palestine from 1882 onwards also signalled a gradual change in Jewish attitudes towards Arabic that gained pace as the national conflict in the country heated up. The shift, as demonstrated in Mendel (2014), went from studying Arabic for daily communication, desire for social integration or financial considerations to a different kind of interest in the language: one that was connected to the political aspirations and the perceived security needs of the Zionist leadership in Palestine.

But in 1882, it was still not possible to foresee the shift that was about to take place. The first Zionist immigrants had varied approaches to Arabic and, generally speaking, these early groups did not have a unified idea about the local Arab-Palestinians, nor about their language. Therefore, their perceptions of the local population and language were mixed and ranged from admiration to disgust: some found positive aspects to the study of Arabic while others intentionally ignored the language and considered it unworthy of their attention.

One of the first groups to come to Palestine was the *Bilu*, a small group of Ashkenazi secular Russian Zionists, who escaped from anti-Jewish pogroms and dreamed of establishing a Jewish home in Erets Yisraeli. Despite the small size of the group – at its height the group consisted of a few dozen members in Palestine – their perceptions of Arab-Palestinians and their language are significant as they are believed to have planted the first 'seeds' of political Zionism, and they have been portrayed in Zionist mythology as 'pioneers, blazing the path for the Jewish masses that would follow in their footsteps' (Shapira, 1992: 56). On top of this, the Bilu serve as a good example to the Zionist ambivalent attitudes towards the local Arab-Palestinians, a dual sentiment that accompanied later stages of Zionist institutions and thought.

The first Bilus arrived in Palestine in the summer of 1882, and in their desire to become locals in Palestine and to shake off their diaspora characteristics, they imitated and romanticised the Arab-Palestinians. This included working in the fields like the local *fallāḥs* (peasants) wearing black and white *kūfiyyā*s on their heads, and also a gradual desire to chat in Arabic. This was one important adjustment made by these Ashkenazi Jews, and Almog (2000) highlights the importance that Arabic had at the time as a symbol of Zionist nativism. One way or another, these immigrants believed their 'return' to Erets Yisraeli was a chance for a historic meeting with 'the Arabs', not as enemies but as 'two brothers of the Semitic family, the children of Abraham' (quoted in Elboim-Dror, 1990: 129). In this view, the Palestinian *fallāḥ*s or local Bedouins were seen not as the ultimate 'other' but as a romantic reflection of the ancient biblical Jewish 'self'.[20]

'Romantic' could also describe the way Theodore Herzl, the founder of Political Zionism, saw the local Arab-Palestinians, and their language. In his utopian novel, *Altneuland* (The Old-New Land), published in 1902, Herzl outlined his vision for a Jewish polity in Erets Yisraeli. In his vision, Zionist immigration to Palestine would prove beneficial for the local population, though the potential Palestinian-Zionist national tension was highlighted as a possible obstacle to Zionist aspirations. The view Herzl imagined future Arab-Palestinians would have of Zionism was provided through the fictional character of Rashid Bey, an educated Arab-Palestinian. According to Rashid Bey, the Arab-Palestinians were 'not at all angry about the increased Jewish immigration to the country'. 'What kind of stupid ideas do you have in your mind?' Bey replied in fluent German – the language Herzl envisioned would be spoken in the future Zionist Palestine – to a person who asked him about Zionist immigration, 'Would you call someone who never took anything from you, and kept giving you more and more, a thief?!' (Herzl, 1997: 98–9).

Interestingly, Herzl does not mention in his book that the local Arab-Palestinians spoke Arabic, but in reality the practical question of studying the language gained importance with the early immigrants. Some of the Zionist groups, including the Bilu, made an effort to learn the language independently. These efforts resulted from the ambivalent thinking that seems to have shaped mainstream Zionist thought in the pre-state period: it was a mixture of admiration for the Arab *fallāḥ*s (although rarely for the urban Arab-Palestinians); a wish to be connected to the land like them; but also a desire – hidden or overt – to take their place. This desire can be seen, for example, in the idea of 'creating' a Jewish peasant who works and looks like the Arab *fallāḥ*, but who would be provided with 'European wages' through *subsidisation* from Zionist institutions in order to exclude the Arab *fallāḥ*.[21] This phenomenon can be observed in the actions of two small Zionist groups – *Ha-Shomer* ('the Watchman') and *Ha-*

---

20   Some of the Zionist immigrants actually regarded the Palestinian *fallāḥ*s as Jews who had remained in the land after the destruction of the Jewish Temple in Jerusalem in 70 AD, and who later converted to Islam. This view was later expressed by Israel's first Prime Minister, David Ben-Gurion, who argued that the Palestinian *fallāḥ*s were Jewish farmers who, in the hardest times, preferred changing their religion to leaving their land. See: Gil Eyal, *The Disenchantment of the Orient: Expertise in Arab Affairs and the Israeli State* (Stanford, CA: Stanford University Press, 2006), p. 53.

21   On the connection between subsidisation and the Zionist aim of exclusion of Arab workers, see Gershon Shafir, *Land, Labor and the Origins of the Israeli-Palestinian Conflict, 1882–1914* (Berkeley, CA: University of California Press, 1996), p. 61.

Ro'eh ('the Shepherd') – whose members came in the second wave of Zionist immigration to Palestine, and who are important for our understanding of the evolvement of this ambivalence.

The origins of Ha-Shomer (established in 1909), which at its height numbered over a hundred members, go back to 1907 when a group of Jewish activists from the *Po'aley Tsiyon* ('Workers of Zion') political party created a Jewish watchmen organisation titled *Bar-Giyora* whose main aim was to provide guard services to the new Jewish settlements. Interestingly, Bar-Giyora was named after Simon Bar-Giyora, a Jewish military leader who fought against the Romans in Jerusalem in 69 CE. The name of the new group encapsulated the Zionist ideal of 'returning' to 'their' land and a desire to emulate the powerful and brave Jewish historical figure through the creation of the 'new Jew'. Another interesting point, which illustrates the 'admiration and exclusion' mentioned earlier, was the context for the creation of Bar-Giyora. The 1907 decision was taken following a couple of decades during which local Arab-Palestinians were mostly hired for the mission of guarding Jewish settlements.

Ha-Shomer's acceptance tests, following those of Bar-Giyora's, required proficiency in the use of guns, horsemanship and *knowledge of Arabic* (Gera, 1985). The members of this organisation, who were in contact with local Arab-Palestinians, had an ambivalent attitude towards the latter: they admired them, sometimes even eroticised them, but they also wanted to usurp them. Ha-Shomer member Zvi Nadav's fantasy was 'to marry four Bedouin women, who will give birth to many strong and healthy boys' (Kimmerling, 2004: 127). This serves as a glimpse into the world of Ha-Shomer, in which Oriental eroticism, admiration and perhaps also the notion of 'improving' Jewish national and physical strength were intertwined.

The selected watchmen who successfully finished their training were hired to protect the new Jewish settlements, a job that had previously been done by Arab-Palestinians. Therefore, the fact that the Jewish watchmen rode horses, spoke Arabic, and wore *kūfiyyā*s (Arab headdress) and *jallābiyya*s (an Arab dress) should not be analysed as only romantic imitation (see figures 1.1, 1.2 and 1.3). Aspects of admiration of the Arab's appearance and a wish to be *like* him were indeed present, but more than this, the watchmen wanted to be a perfect Jewish *replacement* for the Arab.[22] In other words, in the act of becoming a Jewish watchman, one had to go through a period of 'Arabicisation' and learning from

---

22　This corresponds with Hever's literary analysis regarding Zionist literature and the place of the Arab within it. Hever points out that, in the literary sphere, there was no denial of the Palestinian 'other' but an attempt to situate him in an inferior position. Through what Hever depicted as a 'masking mechanism', Zionist literature strives to depict a Palestinian 'other' who is similar to the Zionist 'self' but less ominous, and also to focus on the powerfulness of the 'self' in order to ensure the subordination of the 'other'. See: Hannan Hever, 'Territoriality and Otherness in Hebrew Literature of the War of Independence', in Laurence Jay Silberstein and Robert L. Cohn (eds), *The Other in Jewish Thought and*

the local Arab-Palestinian, before it was possible to replace him and complete the Jewish 'return'.

Likewise, Ha-Ro'eh ('the Shepherd') had its own romantic Orientalist approach. This group, created in 1914, was one of the notable offshoots of Ha-Shomer, and its members had to learn the art of shepherding from local Bedouins. Gershon Fleischer, one of the members of Ha-Ro'eh, remembers that the movement was created when members of Ha-Shomer concluded that in order to improve their ways, they 'must be mobile like the Bedouins, live in tents like them, herd sheep and cattle like them. ... We decided to go to the Bedouins and learn the art of shepherding from them' (Soshuk and Eisenberg, 1984: 75–9).[23] Indeed, during their apprenticeship period, wearing a long Bedouin sword under their white *jallābiyya*s, the Jewish-shepherds learned from their Bedouin 'masters' the Arabic language, as well as how to herd, raise and tend to sheep, and how to make a tent. They also accustomed themselves to goat milk, curds and the difficult outdoor living conditions (Kimmerling, 2004).

In short, there was a strong and clear emphasis in these pioneering organisations on becoming 'local' and 'native', which overall necessitated the imitation and adaptation of Arab-Palestinian local customs, and learning the Arabic language. On top of this, it was in these groups that the process of the creation of a Jewish national character was made possible through Arabisation and de-Arabisation. However, and this is the important principle, when this period of practice ended, members of these groups returned to their settlements and gradually *replaced* the Palestinian shepherds and guards they had previously hired and learned from.

By imitating some of the indigenous population's customs, the Zionist pioneers created the desired 'new Jew': a Jewish identity that differentiated itself from its European past, that had local Oriental characteristics, but that was not integrated or identical to the local Arab-Palestinians. This phenomenon was connected to the important Zionist concept of promoting Jewish labour in Palestine, which was to be known as 'Conquest of Labour' or as *Ḵibbush ha-'Avodah*.[24] The creation

---

History: Constructions of Jewish Culture and Identity (New York: New York University Press, 1994), pp. 242–3.

23   A number of writers have highlighted the importance of the Bedouin image to the early construction of the Zionist-settler identity. For example, Bartal draws a comparison between the image of the Cossack, which was also imitated in the early Zionist period, and the image of the Bedouin, as two images of the 'Other' which evoked both fear and the desire to be like them. The identification with an intimidating figure helped to lessen the fears of Zionist settlers and helped them to become more independent and more part of the land, even if they felt they lived amidst a hostile environment. See: Israel Bartal [in Hebrew], *Cossack and Bedouin: Land and People in Jewish Nationalism* (Tel Aviv, Am Oved Publishers, 2007).

24   For further reading about the concept of Jewish Labour in Zionism, see Zachary Lockman, *Comrades and Enemies: Arab and Jewish Workers in Palestine, 1906–1948* (Berkeley, CA: University of California Press, 1996), pp. 50–53; Zachary Lockman (1993)

Figure 1.1   Three Boys Dressed up as Ha-Shomer Guards (late 1930s)
Source: Courtesy of the Rehovot Archive.

Figure 1.2   Members of Ha-Shomer 1909
Source: Courtesy of the Central Zionist Archive.

קבוצת רוכבים מ."השומר" בגליל התחתון

Figure 1.3   Ha-Shomer Riders in Lower Galilee
Source: Courtesy of the Central Zionist Archive.

of a 'new' *Jewish* shepherd, therefore, or a 'new' *Jewish* watchman, served as an example of, and explanation for the Zionists' general aim of enrooting in the country on the one hand, and segregating the Jewish community from the local Arab-Palestinian population on the other. Imitation, therefore, can be read as an intermediary step towards separation and replacement, since it first allowed the new Jewish immigrants to gain the required skills, but then led to the creation of a Jewish alternative.

The linguistic aspect that this chapter is focused on is related to the same process described above and seems to complement and correspond well with this notion. Arabic, as mentioned before, was needed by the 'new Jew' in Palestine, even on a basic level. However, it was Hebrew, and its revival, that became the symbol of the Jewish 'return' to the Orient. From the last decade of the nineteenth century, the modern revival of Hebrew gained in pace and popularity and was led by, among others, Eli'ezer Ben-Yehouda. The 'revival of Hebrew' played a prominent part in Jewish national aspirations, as well as in the desired transfer from the old Jewish European diaspora identity associated with German, Yiddish or Polish, to the status of the 'new' Hebrew person (Spolsky and Shohamy, 1999).

Interestingly, in a similar way to the acts of Ha-Shomer and Ha-Ro'eh, Ben-Yehouda found great significance in the study of Arabic as the basis for the 'return' and for 'reviving Hebrew', and in a metaphorical way for the replacement of Arabic. Ben-Yehouda noted that Arabic had a rich and ancient vocabulary, unlike Hebrew, which was 'dead' as a spoken vernacular, and did not have the same continuity. Therefore, and since both languages are Semitic, Ben-Yehouda perceived Arabic as a reservoir of words which could be used in the revival of modern Hebrew. His explanation is based on the similarity of the root system that exists in both Arabic and Hebrew.[25] Interestingly, by referring to the similarities between the structures of the two languages, Ben-Yehouda linked the Hebrew *linguistic root system* to the *national roots* of Zionism in Palestine. As Ben-Yehouda (1912: 9) states:

> Only those who, like me, compare words between these two languages [Arabic and Hebrew] can feel how little difference there is between them. You can actually decide that every root in Arabic also exists in Hebrew ... We are allowed to rule that most of the words found in the Arabic vocabulary also existed in the Hebrew vocabulary, and that therefore these are not foreign roots. They are even

---

'Railway Workers and Relational History: Arabs and Jews in British-Ruled Palestine', *Society for Comparative Study of Society and History* 35(3): 608–9.

25   According to the Semitic linguistic system, the root of each verb consists of two to four letters, and can be conjugated according to different forms called *binyanim* in Hebrew, or *awzān* in Arabic.

not Arabic roots. They are our roots. They are the roots that we lost, and that now
we are coming back to find them.[26]

This fascinating approach draws similarities between the processes of 'imitation
and replacement' vis-à-vis the training of members of Ha-Shomer and Ha-Ro'eh,
to that of the 'revival of Hebrew'. It shows how in these cases the connection to
Arab customs and culture was made on the basis of a Jewish 'return' as well as on
principles of separation, all sharing an almost military 'conquest' element in Zionist
discourse: from *Kibbush ha-'Avodah* (Conquest of Labour) through *Kibbush ha-
Adamah* (Conquest of Land) to *Kibbush ha-Safah* (Conquest of Language).

These 'conquests' corresponded with socio-political Zionist justifications
and rationalisations for the displacement of the Palestinians (Mendel, 2014:
25). According to this line of thought, the Zionist movement needed to be
'grateful' to the local Arab-Palestinians who fulfilled their 'historical role' by
preserving the alleged biblical Jewish customs, place names and values that
were all 'waiting' to be 'redeemed' or at least 'relearned' and reconstructed by
the Zionist movement.[27] This moment of hybridisation, as a result, was not a
genuine attempt at integration or coexistence. As the Jewish national movement
became stronger its contact with the local population continued in tandem with
its emphasis on replacing and removing them.

Arabic and the Zionist Movement: Language of the 'Self' vs. Language of the 'Other'

During the same period, the question of Arabic studies was discussed at the
highest levels of the Zionist leadership. One of the main supporters of its
inclusion was the philologist Yitzhak Epstein, considered one of the first
Zionist thinkers of the period to deal publicly with the question of Jewish-
Arab relations in Palestine. In 1905, together with Dr Yosef Lurya, he spoke

---

26   The idea of a commonality was also apparent in a number of common descent
theories that were prevalent in the early twentieth century. According to Zerubavel, because
of their perceived traditional and immemorial lifestyle, the *fallāh* and the Bedouin offered
Zionist settlers an imagined 'symbolic bridge' to ancient Hebrews and the Bible. Some
scholars even went so far as to claim that they were descendants of Jewish farmers who had
stayed behind after the exile and converted to Islam. See Yael Zerubavel, (2008) 'Memory,
the Rebirth of the Native, and the "Hebrew Bedouin" Identity', *Social Research* 75(1):
315–52.

27   For example, the Names Committee of the Jewish National Fund ruled in 1949 that
if an Arab village had a name that was probably a 'disruption of the original Jewish name',
its name needed to be 'changed back to the historical-Hebrew *original* one'. See: Noga
Kadman [in Hebrew], *On the Side of the Road and in the Margins of Consciousness: The
Depopulated Palestinian Villages of 1948 in the Israeli Discourse* (Jerusalem: November
Books, 2008), p. 58.

at the Jewish Congress in Basel about the need to study Arabic as a means of rapprochement between Jews and Arabs, and as a way of learning Arab customs and culture (Snir, 2006). In 1907, Epstein published an article called 'The Forgotten Question' in the Jewish journal *Ha-Shilo'aḥ*. According to him (quoted in Elbonim-Dror, 1990: 360):

> We have forgotten that there is in our beloved land an entire nation that has been there for hundreds of years and who never intended to leave. ... We must get to know the Arab people properly, their ambitions and literature. ... It is a disgrace that nothing has been done on that matter yet.[28]

Similar views were voiced by Dr Nissim Malul, a Palestinian-born Sephardic Jew, who later played an important role in the Zionist Arabic publications project. In an article published in *Ha-Ḥerut* ('The Freedom') newspaper in 1913, he argued that integration in Arab culture is essential for the revival of a Hebrew culture. In another article entitled 'Our Position in the Country', he relied on his great knowledge of the Arabic language and culture and stated that 'we must consolidate our Semitic nationality and not obfuscate it with European culture. Through Arabic we can create a true Hebrew culture' (quoted in Gorni, 1987: 48).

However, despite Malul's, Lurya's and Epstein's desire to become acquainted with the Arabic language, they faced strong opposition. Many other Zionist leaders thought that there was no need to familiarise themselves with either the Arab-Palestinians or their language. Scholars such as Asher Ginzburg and Yosef Klausner believed that 'in studying Arabic there is an unnecessary Levantinisation of the Jewish people' (quoted in, Spolsky and Shohamy, 1999: 139). This contempt towards the local Arab-Palestinians and their culture was not rare, and some of the most important Zionist leaders expressed similar patronising attitudes towards them.

It is not entirely clear, therefore, whether the trend for teaching Arabic in Jewish schools in Palestine, which gained momentum at the beginning of the twentieth century, stemmed from a mostly integrative or instrumental orientation, to use Gardner and Lambert's (1972) concept. Either way, in the course of time, and for several reasons – among them the practical need to know Arabic, pressures from parents to teach the language, the establishment of new European schools in Palestine, or the belief that studying Arabic would support the study of Hebrew – about 5–6 hours of Arabic a week were included in the curriculum in the majority of Jewish schools in Palestine from 1911 onwards (Moreh, 2001).

Interestingly, in 1917 the collapse of the Ottoman Empire led to the disappearance of Ottoman Turkish from all institutions, which in turn enhanced

---

28   See also: Alan Dowty (2001) '"A Question that Outweighs all Others": Yitzhak Epstein and Zionist Recognition of the Arab Issue', *Israel Studies* 6(1): 34–54.

the status of Arabic in Jewish schools in Palestine and reinforced it as the second foreign language, after Hebrew and English. This resulted from the establishment of British rule in Palestine, which signalled the beginning of a period of increased institutionalisation and standardisation in all aspects of social life, and which serves as another explanation for the increased discussions within the Zionist movement on the need for Arabic studies (Halperin, 2006). During that period, and parallel to the intensifying conflict between the Jewish and the Arab-Palestinian communities, two somewhat contradictory aims of Arabic studies were formulated by the Zionist movement: 'getting to know' the language of the *neighbour* in order to conduct peaceful relations, and studying the language of the *enemy* in order to fulfil Zionist security needs as well as its political separatist desires.

In other words, the mixture of romanticisation, admiration and imitation of the local Arab-Palestinian, together with a desire (overt or covert) to become a political replacement for this people, were all evident throughout the Zionist immigrations to Palestine. One can argue that only the *scale* of each process changed with time: that, as the conflict intensified, the percentage of Jews in Palestine grew larger and the idea of creating a Jewish national home in Palestine became more real, the 'volume' of admiration for Arab-Palestinians among Zionists decreased and that of the desire to replace them increased.

This duality can clearly be seen when comparing the first and second *'aliya*s (Jewish immigration waves). First *'aliya* (1882–1903) immigrants operated in a similar fashion to traditional colonial plantation owners. They employed a mostly Arab-Palestinian and Bedouin workforce, including women as maids and nannies, and traded and bought local food and building materials as a way of acclimatising, connecting to the land and with their neighbours (Bartal, 1976). However, they rarely emphasised or discussed the fact that their acts might impose a political threat to the Arab-Palestinian majority. Second *'aliya* immigrants (1904–1914) continued to romanticise the local Arab-Palestinians while working to establish a Jewish-only economy, in this way trying to become a perfect *substitute* for the local Arab-Palestinian.

The Jewish pre-state school system provides another example of the way romanticising and appreciating Arabic was replaced with 'foreignising' it. Looking at the example of the Hebrew Reali School in Haifa, the leading school in the field of Arabic studies at the time (end of the Ottoman Empire and throughout the British Mandate in Palestine), one can better understand how the political situation 'dictated' a new, less integrative approach, to Arabic.[29] It is clear that the Hebrew Reali School had a significant pioneering role in the Jewish education system generally, and in the field of Arabic studies more specifically. Arthur Biram, the school's principal, was a German-Jewish pedagogue for

---

29   For the unique case study of the teaching of Arabic in the Hebrew Reali School, see: Yonatan Mendel, 'From German Philology to Local Usability: The Emergence of "practical" Arabic in the Hebrew Reali School in Haifa – 1913–1948', *Middle Eastern Studies* (forthcoming).

whom the idea of teaching Arabic was an essential part of running a Hebrew School. The same approach was taken by other German-Jewish philologists who operated in the school, most famously S.D. Goitein and Martin Plessner. The interesting point to make here is that from the school's establishment (1914) and until the beginning of the Great Arab Revolt (1936), the attitude towards Arabic was both positive and integrative. It was seen as part of improving the knowledge of Hebrew and getting to know the region. These ideas, which stemmed directly from the German Oriental approach, connected grammar studies, classical texts, disciplinary virtues and Jewish-Muslim interactions. Biram argued that, through Arabic studies, pupils would be able to learn the compositions and creations of Jewish philosophers and intellectuals who worked in the Islamic and Arab world, especially during medieval times. Hence, according to him, through Arabic, the pupils would become acquainted with Jewish-Muslim integrations, with humanistic values that were produced at the time, and with cultural values that prospered within the Muslim societies in which Jewish thinkers operated (Halperin, 2006; Milson, 1996).

In parallel to this, Biram's approach towards Arabic studies also included another German-oriented aspect, which presented Arabic as the *Latin of the Middle East.*[30] According to this notion, the study of Arabic grammar, with its logical set of linguistic rules, would have a positive and constructive effect on formal education. In that regard, the teaching of the grammatical Arabic concept of *I'rāb* (inflexion) was comparable to the teaching of Latin *casus* and, more generally, the importance of Latin for students in Germany was compared with the importance of Arabic for Hebrew students in Palestine. This, according to Biram, would result in disciplinary values connected to the study of grammar, which would be based on a comparison between the virtues of Latin for European schools and the virtues of Arabic for Jewish schools in Palestine,[31] and would improve the pupils' precision of thought. 'Arabic should become the Latin of the Orient!' Biram used to declare in the Reali, emphasising the importance of proper and compulsory teaching of Arabic grammar in the school (Halperin, 2006).

Lastly, Biram supported the study of Arabic as it was linked, for him, with the study of Hebrew. Biram repeatedly highlighted the importance of the study of Hebrew through Arabic, and of the use of Arabic as an explanation for grammar and syntax difficulties that arose during the study of Hebrew. Biram believed that this kind of study would help to develop a 'Semitic language sensation' that would be needed for the study of both Hebrew and Arabic, altogether bringing them closer together, in sound and origin (Milson, 1996: 177). This meant that,

---

30 On the perception of Arabic as the 'Latin of the East', see: G. Dagmar and W. Reuschel, 'Status Types and Status Changes in the Arabic Language', in U. Ammon and M. Hellinger (eds) *Status Change of Languages* (Berlin: de Gruyter, 1992), pp. 85–6, 94.

31 Mentioned in 'Official Report: 1928–1929 – The Hebrew Reali School' [in Hebrew], p. 9, The Aviezer Yellin Archives of Jewish Education in Israel and the Diaspora – Tel Aviv University, 8.45/3236.

through the study of Arabic, Jewish pupils would be able to 'return' to their Semitic roots, on both spiritual and physical levels, and that Arabic would serve as fertile ground for the 'awakening' of Hebrew – in a similar way to Eli'ezer Ben-Yehouda's approach.

The same ideas can be seen in the first Arabic textbook written for Jewish students in the Hebrew education system in Palestine. The two scholars who wrote it, Levy Billig and Avino'am Yellin, completed work on the various Arabic manuscripts for the Arabic Reader they co-edited in 1930. It was entitled *Collections of Readings* (lit. *Liḳuṭey Ḳri'a*), and was published in 1931. Titled in Arabic *Mukhtārāt al-Qirā'a*, the two editors mentioned in their introduction that the texts compiled in the book were chosen 'according to their cultural value in the life of Arab people, and their general cultural value ... as well as *Arabic literature that connects them to the history of the people of Israel*'. They also highlighted the direct support and encouragement of Biram in the making of the book (Billig and Yellin, 1931).

The Arabic Reader, the first ever educational material designed for Hebrew students, brought together classic compositions from the history of Arabic writings, and included works from the pre-Islamic era, through the period of the Prophet Muhammad and the era of the first Caliphs, to the Umayyad period and the late period. The variety of works in the anthology indicate the knowledge that Billig and Yellin wanted Jewish students of Arabic to acquire, which was a profound combination of classical Arabic writings from all times. Works such as those of Albukhārī, Al-Ṭabarī and Al-Shahristānī introduced the student to the profundity of Islamic and Arab thought, and other works – such as the travelogue of the great fourteenth-century Arab explorer Ibn Baṭūṭa – aspired to bring the Jewish students closer to Judaism, Erets Yisraeli and the Arab people. This is clearly seen in their decision to include Ibn Baṭūṭa's journey to the 'country of Palestine' (lit. *Bilād Filasṭīn*), which described a visit to the city of 'Asqilān (Ashkelon), and the towns of Al-Ramlah (Ramla), Nablus, and 'Akka (Acre). Texts such as this were included in order to bring the Jewish pupils closer to the region on a historical, geographical and cultural level.

Interestingly, the second Arabic textbook designed for Jewish students in Hebrew schools connected Arabic to Hebrew (and Judaism) even more clearly. Completed in 1935, and entitled *The Theory of Arabic Grammar: A Guidebook for Hebrew Schools* (lit. *Torat ha-Diḳduḳ ha-'Aravi: Sefer 'Ezrah le-Vatei Sefer 'Ivriyyim*), the textbook was composed by Martin Plessner. In the explanation of the book's rationale, Plessner emphasised the unprecedented character of the act of composing this kind of Arabic grammar textbook designed for Jewish pupils. 'I would like my readers and critics to remember', he wrote in the preface, 'that I did not have any similar example to follow' (lit. *'lo haya li mofet'*). Plessner meant that, in making Arabic grammar studies coherent for Hebrew students, he had to find ways to connect and clarify similarities between the two Semitic languages. This, he explained, was done:

> as I wish the study of Arabic grammar to bring the [Jewish] students to even deeper understanding of Hebrew ... through focusing on the systematic grammatical structures of Arabic ... and through highlighting how the language of the Bible can be better understood through the study of Arabic. (1935: 1)

However, this educational approach to Arabic did not last. From the 1930s, and especially after the 'Great Arab Revolt' in 1936, the general attitude shifted from one that viewed Arabic as a source for humanistic knowledge and for highlighting Jewish-Muslim shared history and values, to one that was 'practical' and focused on instrumentality. This was a result of the changing political situation, and a gradual tendency to place more emphasis on political and current affairs. This changing balance shows, at the end of the day, the study of Arabic did not bring the Jewish and Arab communities in Palestine closer together. It also failed to instil the idea of Arabic as a sister of Hebrew and as an important mark of identity for Jewish students. It did quite the reverse. As such, and as delineated by Raz-Krakotzkin (2005: 166), the field of Arabic studies was part and parcel of the emerging Zionist thought that did not challenge the dichotomy between Europe and the Orient and only corresponded to the desire to be in the Orient physically in a way that would help it to further assimilate into the West.

The Revitalisation of Hebrew

The gradual shift in attitude towards Arabic was clearly seen in the debates of the Committee for the Hebrew Language, established by Eliʿezer Ben-Yehouda at the end of the nineteenth century in Palestine. Iair Or highlights in his research (2015) the strong presence of Arabic in the debates on the revival of Hebrew in the committee.[32] In contrast, it is interesting to see the committee's early approach towards Yiddish – which was seen as the most evident 'enemy' of Hebrew, representing the language of the Jewish diaspora (Chaver, 2004). Alongside the anti-Yiddish campaign, Arabic enjoyed a different status, and was seen as a tool to 'revive' Hebrew (as mentioned before with regard to Ben-Yehouda's quote from 1912). Eisenstadt (2002) exemplified the perception of the committee regarding Arabic and Arabic pronunciation in the analysis of the debates on the 'correct pronunciation' in Hebrew, and the final decision to use the Mizrahi-Arabic accentuation (that puts the stress on the end of the word – *milra*ʾ), unlike the Ashkenazi pronunciation (that puts the stress on the beginning of the word –

32　Compare with Lital Levy's research and insights, and her point that 'an early twentieth-century ethnographer with a camera would have documented not the influence of Hebrew on Israeli Arabic, but the influence of Arabic on Palestinian Hebrew'. In: Lital Levy, *Poetic Tresspass: Writing Between Hebrew and Arabic in Israel/Palestine* (Princeton: Princeton University Press, 2014), p. 26.

mil 'el).[33] The Mirzahi-Arabic pronunciation was considered to be more *ancient*, more *authentic, natural* and *local*, and representative of a Jewish acceptance into the Muslim-Arab area, while the Ashkenazi pronunciation represented the diasporic Jewish life. When Ben-Yehouda had to rule on whether to use the Mizrahi-Arabic pronunciation or the Ashkenazi-European one, he said:

> the pronunciation of the Mizrahi Jews is so original, so Oriental and sweet. The Mizrahi sounds of ṭet, 'ayn and quf is like music to my ears. Their pronunciation is so natural. Their image is so beautiful. ... While the Ashkenazi pronunciation lacks this music and lacks this Mizrahi courage.

This was the original view of the committee and, based on this perception – and the previous idea of returning to Hebrew while *borrowing back* words and roots from Arabic – the Arabic language was used as a reservoir for the 'revival' of Hebrew'. When Ben-Yehouda refers to the process of enriching modern Hebrew with Arabic words, he writes in his life-long project *The Hebrew Dictionary of the Hebrew Language: Old and New* (lit. *Milon Ben Yehouda la-Lashon ha-'Ivrit ha-Yeshana ye-ha-Ḥadasha*) that Arabic was constantly used in the process of 'reviving Hebrew'. According to Ben-Yehouda (1980: 10):

> I tended to conceptualise the Hebrew verbs in relation to the Arabic verbs, first as Arabic serves as a witness for a language that has ceased to exist. ... and, second, as these similarities will help the speakers of Hebrew to realise how close these two languages are, in their spirit and characteristics, so close they are actually one language.

Elsewhere Ben-Yehouda wrote:

> when we tried to create a noun out of a verb that was known to us, I used to look at the noun's name in Arabic, which is Hebrew's sister language (lit. *laḳaḥti li et leshon 'Aravit aḥota le-mofet*), and used it as my example for the creation of the new Hebrew noun. (1980: 13)

Following this, the Committee for the Hebrew Language concluded in 1929 (1929: 16) that when 'renewing' a Hebrew verb 'it is recommended to use the Semitic verbs: Aramaic, Canaanite, Egyptian and especially Arab roots'. Amit-Kochavi (2011: 1590) highlighted this phenomenon in Ben-Yehouda's project, stating that as he saw the two languages almost as one, he 'freely borrowed and adapted Arabic roots and vocabulary to coin new Hebrew words'. As examples,

---

33   For further reading, see: Shmuel Noah Eisenstadt [in Hebrew], *Jewish Civilization: The Jewish Historical Experience in a Comparative Perspective and its Manifestations in Israeli Society* (Beersheba: The Ben-Gurion Heritage Institute – Ben-Gurion University of the Negev, 2002).

these words included, among many others, the Hebrew *adiv* (lit. polite) that stemmed from Arabic *adīb*, the Hebrew *rishmi* (lit. official) from the Arabic *rasmī*, the Hebrew *mivrak* (lit. telegraph) from the Arabic *baraqiyya*, the Hebrew *timrun* (lit. manoeuvre) from the Arabic *tamrīn*.

In addition to these examples, perhaps the most well-known Arabic influence on modern Hebrew, is the use of the suffix *–iyyā*, which through the Arabic impact has become part and parcel of modern Hebrew. This, for example, was the case of the *'iriyya* (lit. municipality) which is based on the replacement of the Arabic word *balad*, meaning 'city' (as in *baladiyyā*, municipality in Arabic) with the Hebrew word *'ir* which means 'city' in Hebrew; *yamiyya* (lit. navy) which is based on the replacement of the Arabic word *bahr*, meaning 'sea' (as in *bahriyyā*, navy in Arabic) with the Hebrew word *yam* which means 'sea' in Hebrew; and *shimshiyya* (lit. parasol) which is based on the replacement of the Arabic word *shams*, meaning 'sun' (as in *shamsiyyā*, parasol in Arabic) with the Hebrew word *shemesh* which means 'sun' in Hebrew (Blau, 1976: 87).

It is important to note that the influence of Arabic on Hebrew extends beyond the modern period. As Blau (1981) notes, during the medieval period, Arabic also played an important role vis-à-vis Hebrew. However, Blau also argued that the influence of modern Arabic on modern Hebrew (as is the case with Ben-Yehouda) was much greater than the influence of Medieval Arabic on Medieval Hebrew.

The influences of Arabic on the revival of Hebrew soon hit the wall of the conflict in Palestine. As the conflict intensified, the debates in the Committee for the Hebrew Language were not as supportive of the 'return' to Hebrew via the 'return' to Arabic. Instead, a strong desire to distinguish between Jews and Arabs, Arab labour and Jewish labour, Arab land and Jewish land, and also Hebrew and Arabic, was expressed. Shenhav (2012) highlights how, in the 1930s, Arabic became strongly associated with 'the language of the Arab enemy'. In light of this perception, the committee began to prefer to base new Hebrew words on biblical Hebrew (the alleged 'purest' Hebrew) or on the Hebrew of the *Talmud* and *Mishna* (Jewish rabbinic texts that included obvious influences of Aramaic and Arabic). Shenhav quoted Avinery (1964: 457), who argued that, at that point, Arabic ceased to be a desirable language, and that new:

> borders appeared, borders that if one cross them there might be a deviation from the national [Jewish] framework into other [Arab] nations. … With relation to Arabic the two people [Jews and Arabs] are far away from having peaceful and real brotherhood … and while material similarity between the languages exist, a similarity between their spirits and soul do not exist whatsoever.

The same was the case with the accent and pronunciation. While at the beginning there was a desire to speak with Mizrahi or Arabic accent in Hebrew, voices in the committee (both because of their Ashkenazi origin and because of the simmering conflict) called for a 'European' version of Hebrew. The 1929 report of the committee stated: 'We come from Europe and the Semitic sounds [in Hebrew] are difficult for us to pronounce', was written in the 1929 report of the committee. Jabotinsky added, in the same spirit, that Hebrew letters should have European – not Arab – sounds. According to him:

> [T]he characteristics of the European Jews do not enable them to speak with an Oriental accent ... especially difficult are the three letters – ṭet, ʿayn and quf ... but who said the letters' [Jewish] ancient pronunciation was similar to the contemporary Arab one? In our new [Europeanised Hebrew] language these letters need to have sounds that fit our musical taste, and this taste is first and foremost European and not Mizrahi. (Shenhav, 2012: 162)

Strikingly, the shift in attitude caused the committee to adapt words into Hebrew that had *as little* connection to Arabic as possible. Arabic was seen as a foreign language that 'in the diaspora' had influenced and distorted Hebrew. Therefore, for example, the decision to prefer the word *geshem* (lit. rain) over the Hebrew biblical word *maṭar* was due to the latter's proximity to Arabic *maṭar*; the decision to prefer the word *karoz* (lit. manifest) over the classical Hebrew word *minshar* was due to the latter's proximity to the Arabic *manshūr*; and the decision to prefer the word *daḳar* (lit. 'to stab') to the biblical word *ṭaʿan* (as in Isaiah 14:19) was due to the latter's proximity to the Arabic *ṭaʿana* (Shenhav, 2012: 163).

Harshav (1990) argues that the final 'product' called modern Hebrew was, therefore, neither here nor there. With regard to pronunciation, for example, Harshav concludes that the Hebrew that was finally accepted was not Ashkenazi [Yiddish-influenced] Hebrew nor Sephardic [Arabic-Mizrahi-influenced] but the lowest common denominator of them both. It is an alleged Sephardic accentuation that was achieved through Ashkenazi filters.

Arabic and Hebrew post-1948

Following the revitalisation of Hebrew, and following the 1947–1948 watershed, the Palestinian Nakba and the creation of Israel, the Arabic language was gradually neglected within Israel. As time passed, Arabic was less and less considered by Jewish-Israelis to be a sister of the Hebrew language and was more associated with Military Intelligence needs and with security-oriented themes more generally (Kraemer, 1993; Mendel, 2014). The Arabic words that have entered Israeli vocabulary since 1948 all seem to be from the world of 'slang' and include mostly swear words and daily expressions (such as *dakhilak*, *maʿa lesh*, *salamtak* or *ṣababa*). These, more than anything else, indicate the current low social capital of Arabic in Israel.

The processes by which Arabic became a low-status language in Israel, a non-Jewish language and non-desired one are depicted in Mendel's work (2013a, 2013b, 2014, 2015a, 2015b). This process is also evident in the distancing that has taken place between Jewish-Israelis and Arabs, Arabic and Arabic connotations, which happened in parallel to the strengthening of the relationship between Arabic studies and security considerations in Israel. Some statistical analysis sheds light on this topic. For example, research conducted in 1988 sought to assess Jewish-Israeli high school students' attitudes towards studying Arabic. Specifically, the researchers inquired as to why students both chose to study Arabic and also elected to undergo rigorous national testing in the topic for their high school matriculation exams. The survey found that 65 per cent of those who chose to study Arabic did not do so to know their neighbours, and definitely not to further their Hebrew, but due to their 'desire to serve in the army in a position which demands knowledge of Arabic'.[34]

A different survey, conducted in 2006, found that 62.9 per cent of Jewish-Israeli students who sat for end of high school matriculation exams in Arabic mentioned a desire to serve in Military Intelligence as their primary motivation. The same survey looked at teacher attitudes and found that 72.8 per cent of the Arabic teachers surveyed believed that a desire to serve in Military Intelligence was a leading factor in their students' decision to choose Arabic. Strikingly, the survey found that 'the wish to serve in Military Intelligence' was consistently the main reason for studying Arabic in Israel, as indicated by students and teachers, in intermediate and in high schools alike (Him Younes and Malka, 2006). These findings correlate with and support additional research which has highlighted the importance of instrumental considerations – including service in Military Intelligence – in motivating Jewish-Israeli students to study Arabic in school (Amara, 2008; Kraemer, 1990).[35]

The attitudes towards Arabic, not as a 'sister language' but as the language of the enemy and considered by Jewish-Israeli students to be 'much more difficult to learn than English', are all ideologically oriented beliefs that stem from the new state of affairs between Jews and Arabs following the creation of Israel. The anti-Arabic stand of Jewish-Israelis, therefore, runs in parallel with the general anti-Arab voices in Israel that try to push forward the notion that the citizenship of Arab-Palestinians is in doubt and could be revoked. These voices are inseparable from general attitudes pervasive in Israeli society in relation to Arabs and the Arab world. In 2003, for example, a government-appointed investigation committee found that on the state level, and historically since 1948, 'the Arab citizens of Israel have lived in a reality in which they have been discriminated against just

34 Quoted in a study titled *Arabic Studies in Hebrew Schools in Israel* [in Hebrew] conducted by Dr Shmu'el Shay, Tamar Tsemah and Haviva Bar and submitted to the Ministry of Education in August 1988. Found in: ISA, PR-10/4461.

35 Most famous is the research of Kraemer, who argued that because of the ongoing conflict, only *instrumental* and *national security* orientations provide the motivation to study the language in schools. Amara, who similarly identified the supremacy of the military in Arabic studies in Israel, noted that Arabic in Israeli-Jewish schools is taught as a security-language and calls for it to be 'civilianised'.

for being Arabs'.[36] A survey conducted in 2006 revealed that about half of the Jewish citizens in Israel believe that the state should encourage Arab emigration. The same survey also investigated how people felt when hearing spoken Arabic. Some 50 per cent of respondents said that hearing Arabic made them fearful, 43 per cent responded that they felt uncomfortable, and 30 per cent said Arabic aroused feelings of hatred within them (Stern, 2007). Another survey, conducted by the Centre for Research on Peace Education, found that of the 1,600 Israeli high school students surveyed, 75 per cent of the Jewish respondents associated Arabs with being 'unclean, uneducated and uncivilised' (Raved, 2007).

As demonstrated here, Jewish-Israeli views of Arabs are, by and large, negative and such attitudes have a profound impact on the way Israeli society relates to its Arab citizens and to the Arabic language. The ideas of Eli'ezer Ben-Yehouda regarding the proximity between Hebrew and Arabic, and Plessner's notion that through the study of Arabic Jewish-Israeli students would get closer to Jewish religious texts, have been overshadowed by the sounds of Arabic swear words, metaphorically and not metaphorically.

These negative perceptions towards Arabic and its speakers – which are pervasive among Jews in Israel – can also be seen in the diminishing Israeli interest in learning Arabic, and in the low number of Arabic speakers among Jewish-Israelis. Klein, for example, writes (2013):

> Today it is hard to find Arabic speakers in Israel who are not Arabs or who were not born in a Muslim country ... *only three per cent of Israeli-born Jews speak Arabic* [emphasis added]. Last year [2012] only some 2,000 Jewish high-school students took the matriculation exam in the language of 20 per cent of their country's residents. The teenagers who took that test in Arabic did not see it as bridge: they saw it as a weapon, and most of them, presumably, were inducted into [Military Intelligence's] Unit 8200.[37]

In examining research and surveys that have taken place over the past three decades, one could get the impression that animosity between Jews and Arabs, or Hebrew and Arabic, has always been the case. This reading of Jewish Arab relationships has made it hard, not to say impossible, for Jewish-Israeli society to admit that their national language, Hebrew, is very much a language that encapsulates in itself the currently most hated language in Israeli society: Arabic.

---

36   See the Or Committee Report, Chapter 1, article 19, August 2003 [in Hebrew]. The Or Committee investigated the clashes between the Israeli police and Palestinian citizens of Israel during demonstrations at the beginning of the Second Intifada in October 2000. The clashes resulted in 13 Palestinian citizens of Israel being shot dead by the police force.

37   Unit 8200 is the Central Collection Unit of the Israeli Intelligence Corps and the biggest unit in the IDF. See: Daniel Byman, *A High Price: The Triumphs and Failures of Israeli Counterterrorism* (Oxford: Oxford University Press, 2011), p. 339.

Chapter 2

What's in a Symbol? Deconstructing Israeli National Symbols

> If we desire to lead many men, we must raise a symbol above their heads.
>
> Theodor Herzl, 1896[1]

Landing at Israel's Ben Gurion Airport, it did not take long for us to come across an example that illustrated the complicated relationship between Israeli and Arab-Palestinian national identities and cultures. After spending only a couple of minutes in Israel, in the airport's long corridor, alongside airport security, escalators and drinking fountains, we found a 10 meter long bill-board style advertisement that included everything we needed to start our investigation into national symbols. The large indoor billboard was an advertisement for Eldan, a local car rental company and one of country's most popular firms. Established in Israel in 1967, it was an excellent representation of the patriotic and national spirit of Israeli society following the war of June that year. Its founder, Yoseph Dahan, who served as an officer in the Israeli Defence Force (IDF) ordnance corps, wanted to establish a proud, made-in-Israel, nation-wide car rental company. This is evident in the company's name 'Eldan', which means 'to Dan', after the name of the ancient city of Dan, known in the Bible as the northernmost city of the Kingdom of Israel. Along with its somewhat loaded name, the company's logo – the name 'Eldan' printed in Hebrew and English using blue and white colours (the colours of the Israeli flag), which Dahan chose to underline in order to signify that his company is 'really blue and white'.[2]

The images on that billboard, made by a company that strives to be the 'epitome' of Israeliness, were rather telling. The slogan 'In Israel' was written on the far left side of the advertisement followed by three dots, while 'Rent Blue & White, Rent Eldan' was printed on the far right side of the advertisement. Strikingly, in between these two slogans, were three different animated cars,

---

1   Theodor Herzl, *The Jewish State: An Attempt at a Modern Solution of the Jewish Question* [1896] (London: H. Pordes, 1972).

2   See, for example: 'A Rental Car Blue and White' [in Hebrew], *Maariv*, 14 January 1987 http://jpress.org.il/Olive/APA/NLI_heb/?action=tab&tab=browse&pub= MAR (accessed: 18 October 2014). In 2010, on Israel's Independence Day, as part of a promotional campaign Eldan gave away t-shirts titled 'We Are All Blue and White'. The shirts, strikingly, were made in Turkey. See: Ora Koren [in Hebrew], 'The "Blue and White" Campaign of Eldan on Shirts Made in Turkey', *TheMarker*, 14 April 2010, http:// www.themarker.com/misc/1.575877 (accessed: 18 October 2014).

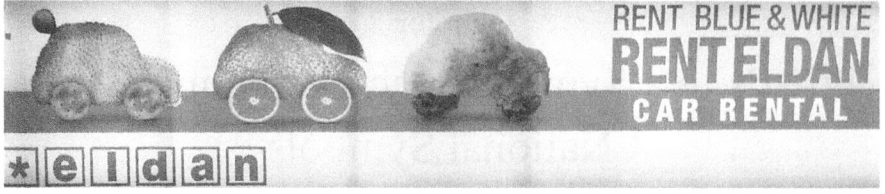

Figure 2.1   Eldan's Car Rental Advertisement

chosen to represent, ostensibly, the most prominent Jewish-Israeli national symbols: the first car had a green body of a cactus and wheels made of prickly pear fruit; the second car had a body of an orange skin and wheels made of orange slices and the third car had a body of a pita bread with falafel balls as wheels (see Figure 2.1). The billboard's obvious symbolism left no room for misunderstanding. The 'Blue and White' was very clearly there as the symbol of Israel's flag and of Israeliness, and alongside it were three prominent pre-1948 Arab-Palestinian symbols: the prickly pear,[3] the Jaffa orange,[4] and the falafel in a pita.[5] These three stood in between 'in Israel' and 'Blue and White' as three Arab-Palestinian symbols that have become totally Israelised.

In a nutshell, this anecdote exposes the process of localisation that was desired by the Jewish immigrants to Palestine from the end of the nineteenth century, and which remained crucial throughout the creation of an Israeli nation-state in 1948. Although this billboard can simply be read as an advertisement for a car-rental company, it also advertises the idea of a cultural shift, from the Diasporic identity of Israel's first immigrants, to a localised and indigenous Hebrew-Zionist identity – one based partly on imitation and appropriation of the indigenous Arab-Palestinian people and the Arab culture. In light of the heated context of the Arab-Israeli Conflict over the course of the twentieth century, Israeli culture has been

---

3   Meron Benvenisti highlights how Palestinian villages, which were totally demolished and erased in 1948, preserved their outlines thanks to the prickly pear cactus hedges. In: Meron Benvenisti, *Sacred Landscape: The Buried History of the Holy Land since 1948* (Berkeley, CA: University of California Press, 2002), p. 170.

4   According to Migdal and Kimmerling, the oranges that grew in the orchards of Jaffa, well known for their quality, were one of the most important reasons that made Jaffa a prosperous Arab-Palestinian city from the beginning of the twentieth century, and the second largest Palestinian city by the end of the Ottoman period in Palestine. In: Baruch Kimmerling and Joel S. Migdal, *The Palestinian People: A History* (Cambridge, MA: Harvard University Press, 2009), p. 48.

5   The transformation of the falafel, from an Arab dish into a central part of the Israeli experience was discussed in our previous study on the emergence of Israeli food culture. See: Ronald Ranta and Yonatan Mendel, (2014) 'Consuming Palestine: Palestine and Palestinians in Israeli food culture', *Ethnicities* 14(3): 413–16, 424; see also Yael Raviv (2003) 'Falafel: A National Icon' *Gastronomica: the Journal of Food and Culture* 3(3): 20–25.

characterised by the bipolarity of its desire to both invent a unique and localised Israeli identity, and adapt and erase the indigenous Palestinian-Arab culture: a process we describe as de-Arabisation.

This was the point of departure for our journey, which began as soon as we stepped out from Ben Gurion Airport to trace both processes: that of the adaptation of Arab-Palestinian elements into the emerging Israeli identity and the processes by which de-Arabising these elements were made possible. In other words, to see how Arab-Palestinian elements have become central symbols of Israeli national identity and thought; how the symbol of the Arab-Palestinian Jaffa Orange was transformed into something as inconsequential as a cartoon wheel powering a blue and white Israeli rental car.

What are National Symbols?

One could argue that symbols, which are often characterised as being related to the imagination and to myths, are of marginal importance to the real world. How could symbols really explain, relate to, encapsulate, or convey something as complex and emotive as national identities, let alone the tense feeling related to the Arab-Israeli Conflict? However, looking at the emotions generated and the tensions that arise from 'simple' actions and decisions, such as the naming and renaming of streets and the careful calculations that go into deciding who or what should appear on the currency in Israel, it becomes clear that symbols and their meaning carry an importance to the nation's identity. This is because symbols are not value and meaning neutral. Rather, they express social and cultural ideas and explain 'who we are' and 'where we come from', and although these notions are subjective, their interpretation is not arbitrary, but is learned and ingrained and is based on communal history and tradition (Elgenius, 2011).

The first thing to note is that symbolism, or the use of symbols as a means of representing ideas and values, is a universal phenomenon. Symbolism is part of the fundamental mechanisms of forming and maintaining group identities and creating, enforcing and defining group boundaries (Smith, 2009). As such, they have been instrumental in the process of forming and constructing national identities. According to Ozkirimli (2005) there are five dimensions to national discourse: temporal, spatial, cultural, political and self/other. National symbols provide a means of accessing, linking and expressing these dimensions. They root the nation in the present while providing links to the idealised future and the mythical past, to communal memories, distant lands and geographies. On top of this, because nations are normally born out of struggle and conflict, symbols signify the achievements, commitments and aspirations, but also the sacrifices and losses of the nation. They, as we argue, serve as the nation's 'flight recorder' or simply its political-cultural 'black box', and can therefore tell a multi-layered story of the crystallisation of national identity. As such, and as Skey (2013) highlights, symbols provide the means of anchoring and solidifying national identities and narratives, as well as linking them to specific images, location and times, which is particularly useful in times of social transformation and change.

With the advent of the nation-state, symbols have performed an important additional task of connecting the nation to the state and its institutions and spreading and enshrining state and elite power and narratives. However, the process of accepting symbols is directly linked to the values and meanings they create in those receiving the message. Symbols thus straddle the line between common values and elite manipulation and 'serve as markers for [the] collective memory of the nation' while supporting 'the power of the state to define the nation' (Geisler, 2005: xv–xvi). National symbols are part of the constituting foundations and heritage of the nation-state. They are used both for recruiting people and convincing them of the appeal of the nation as well as continually disseminating the national ideas once the nation and the state have been formed. National symbols are so powerful and important that we are often willing to put our lives on the line to protect and honour them. The most important aspect of national symbols therefore is not their aesthetic value or history but their representational power and its effect on the nation (Confino, 1997).

Trevor-Roper (1988) offers an interesting angle on the creation, not to say invention of national symbols. In his classic 'The Invention of Tradition: The Highland Tradition of Scotland', he focuses on the Scottish kilt and tartan, and how each 'belongs' to a different, historical clan. Yet, when he looks deeper into these national symbols, Trevor-Roper comes up with the striking finding, namely that, despite their great importance to modern Scottish national identity, aspirations, and economic interests, neither the kilt nor the tartan are 'Scottish'. Though the kilt and tartan were widely used and were part of the history and traditions of the Scottish highlands, they were both adopted and appropriated from neighbouring cultures: the tartan is Irish-Gaelic and the modern kilt, was invented by an English industrialist in the eighteenth century. In addition, Trevor-Roper highlights the fact that the differentiation of kilt patterns, which are now associated with different Scottish clans, is related to financial initiatives to improve weaving production in difficult times.

Why do we not forget our national identities? It is because we are constantly reminded of them through national symbols (Billig, 1995). These come in a variety of forms: from monuments and paper money to advertisements and food items that are generated by a diverse range of actors. Some of the symbols construct a particular image and or convey narrative of the nation, while other symbols represent bottom-up popular conceptions of the nation and its values and ideals. Through these symbols members learn what it means to be part of the imagined community of the nation (Kolsto, 2006); a process which starts in pre-school and continues throughout life. Through viewing national symbols and relating to the discourses surrounding them, the nation becomes sedimented (Skey, 2013). That is to say, through habit and routine, and normally without much reflection, national symbols and the discourses related to them communicate the idea of the nation and transform it from an intangible and abstract idea to something that is viewed as natural, objective and real. As such, the symbols are transformed from simply being images which represent the nation to becoming embodiments of the nation, including its history, values, boundaries and psyche. As a result, symbols are used as means of banally constructing and reproducing the nation in and throughout daily life (Billig, 1995).

Symbols, however, are not merely used as passive representations of the nation. They are important in rooting and relating to the nation in everyday life. Symbols are not simply viewed, they are also used, either to create and enhance particular discourses relating to the nation or as a means of practising and performing the nation in everyday life and at national events (Edensor, 2002; Elgenius, 2011). One cannot imagine national events, such as memorial and independence days, without the accompaniment of monuments, customs, flags, banners, uniforms, etc.

Israel's most central ceremony, and an important national event, the 'Torch-Lighting Ceremony' on the eve of its Independence Day, provides a good example to illustrate the points made above. The ceremony takes place in Jerusalem, on a hill called Mount Herzl, surrounding the grave of Theodor Herzl. The first part of the ceremony comprises of the lighting of 12 torches, after the 12 biblical tribes of Israel, and aims to draw a direct link between the modern *state of Israel* and the biblical tribes of the *Israelites*. This link provides a line of *continuity* and *return* between religion, history, geography and the modern political nation-state.

Following this part of the ceremony, a group of flag-bearing soldiers (lit. *Ha-Daglanim*), whose number is based on the 'age' of the state, conduct military drilling exercises with their flags: the Israeli flag, the IDF flag and the different flags of the IDF units. The soldiers, standing with their flags in different locations within the Mount Herzl compound, create different shapes that symbolise the theme of each year's ceremony. In 2012, the ceremony theme was 'Water as the Israeli source of life' and the IDF soldiers and officers, using their flags and movement, made the shapes of the Sea of Galilee, a biblical well, a water tower, the turbine wheel, and a water tap. In 2013, the ceremony was dedicated to the 'National heritage' and the soldiers created the shapes of a Torah (Bible) book, the 'Yellow Badge' which Jews in Nazi Germany had to wear, a cruse of oil (symbol of the Maccabees' heroism), scales (symbol of justice), a book, a scientific laboratory, the harp of King David, and the Star of David. All in all, the intertwining of soldiers, biblical Jewish elements, political notions and civilian aspects of life serve as a telling example of the ways in which a ceremony can clearly demonstrate characteristics and pillars of the state. In this case, through the use of symbols, the ceremony highlights the 'fluid' boundaries between past and present, civilian and military and Judaism and Zionism, which are all blurred in an attempt to create a unified image of the state and the nation.

Typology of Zionist Symbols

Tamar Meyer suggests that Zionist symbols should be divided into two main groups: the first representing Jewish religious imagery, such as the Menorah (the seven-branch lampstand), which is Israel's national emblem, and the Star of David, which is the central image in Israel's blue and white national flag; the second group reflecting and representing the achievements of the secular Zionist movement, for example, 'immigration to Palestine, reclaiming, settling, and defending the land, reviving the Hebrew language, and establishing Hebrew culture and industry' (2005: 8). She mentions a number of symbols, such as Jaffa oranges and local native trees (olive, oak and palm) as part of the celebration of Zionism's achievements. Meyer's analysis fits in well with the current scholarship on Israeli national symbols, as well as with the type of ceremonies and national events described above.

While we do not disagree with Meyer's analysis, we feel that a third category of Jewish-Israeli national symbols is required in order to convey the full spectrum of Israeli national symbols. This third category includes all of those symbols which were adopted and later appropriated from the local Arab-Palestinian population and culture. These symbols were not taken directly out of Zionist and/or Jewish religious experience and heritage, but adapted and appropriated from local Arab-Palestinians and transformed into Jewish-Israeli ones. It is important to note, though, that many of these symbols, once they were adopted and appropriated, were validated through a process of reconstructive historical writing. That is to say, many symbols were either given a Jewish and or Zionist identity or were related to biblical and Jewish historical precedents. This, we feel, is an important element in the study of Israeli national symbols that has not been given enough space and discussion.

The first two categories of national symbols helped in the process of recruiting a diverse group of people, who at times shared little in common other than a perception of a collective Jewish history, to the Zionist cause and, more specifically, to immigrate to Palestine and take it upon themselves to help establish a Jewish state there. Our proposed third category of symbols was instrumental in connecting the new Zionist settlers to the land of Palestine and enabled their transformation from immigrants to native population through the marginalisation and removal of the indigenous Arab-Palestinian population. This category is also important because it highlights the interesting ambiguity that existed within the Zionist settler-colonial project towards the Arab-Palestinian local population and towards what the future state should look and be like.

When Theodor Herzl (1902) wrote about his vision of a future Jewish state in Palestine, he juxtaposed it against what he saw as the current state of Palestine. For Herzl, pre-Zionist Palestine was a decaying, desolate and backward place; it was dirty, hopeless and neglected; its lands showed little trace of cultivation, being mostly filled with swamps and sand, and it was 'full of motley Oriental misery' a place where 'poor Turks, dirty Arabs, [and] shy Jews lounged around ...

indolent, beggarly and hopeless' (Herzl, 1997: 42). He described Palestine's Arab inhabitants as blackish brigands and as dirty and poor, and, in contrast, imagined a future Zionist Palestine as a Vienna transplanted into the Orient: the customs and costumes, the monuments and political system, the culture and languages, would all be those of Central and Eastern Europe.

Herzl's vision is clear with regard to the first two categories of national symbols. To begin with, many of the early symbols, such as the flag and the anthem had little to do with the land of Palestine in the nineteenth century. They demonstrated a desire to bring together different segments of European Jewry under the banner of a pseudo-secular national movement. The subsequent symbols represented the achievements of practical and socialist Zionism, while showcasing some of the landscapes of Palestine. These symbols contained mostly strong socialist ideals relating to pioneering, hard work, sacrifices and communal life. These symbols also relate directly to the dominant place of socialist Zionism in general and Mapai, the Zionist workers party – the Jewish community's dominant political party from the 1920s to the 1960s. This was done alongside Socialist Zionism's adoption, adaptation and re-interpretation of many Jewish religious symbols and festivals in the attempt to mobilise Jewish society and transform secular Zionism into a civil religion (Don-Yehiya and Liebman, 1981).

The adoption of Socialist Zionist and Jewish religious symbols reflected, to a large extent, the cultural history and ideological background of many of the early Zionist settlers. These symbols, therefore, stood in contrast to later ones that were adopted from, and based on the encounter with the local Arab-Palestinian population and culture. In other words, the adopted and appropriated Arab-Palestinian symbols demonstrate the enormous cultural and social impact the encounter with Palestine and the Arab-Palestinian people had on early Zionist settlers and subsequently on Zionist-Israeli culture and identity. It also places Zionism in a similar context to other settler-colonial endeavours which transformed the settlers into the native population through the adoption of local customs, traditions and symbols.

Israel's National Symbols

As discussed above, Israel has a number of varied national symbols that we argue can be roughly divided into three categories based on their cultural and ideological influences. One could further divide Israel's national symbols into officially recognised symbols, which have been formally approved by the state and its institutions and are mostly top-down generated, and everyday symbols, which are bottom-up and mostly express the general sentiments of the Jewish population in Israel. It is important to clarify that, even though we use the term 'officially recognised symbols', we do not want to imply that these symbols were all top-down generated. Indeed, as will be demonstrated below, many of these symbols

went through a number of complex processes involving a number of different
actors before becoming 'official'.

Among Israel's officially recognised symbols one can find the national
emblem and flag, which are based on Jewish religious motifs, but also a number
of official flora and fauna, which are said to represent the Zionist-Israeli
identity and its relationship to the land of Israel/Palestine, and in particular to
its Jewish biblical past. Probably the most well known of these symbols is the
tsabar (prickly pear; in English sabra). One cannot overstate the importance of
the tsabar/prickly pear (in Arabic: *ṣabr*) image to Israeli identity and culture:
it is the name given to a Zionist settler born in pre-state Palestine and to a
Jew born in Israel. The connection of the tsabar/prickly pear to Israel is seen
literally everywhere in the state, from names of products and companies, to
books discussing the demise of the Israeli Tsabar,[6] and prominent Israeli TV
and literary images and characters. For example, one of the most beloved
and well-known Israeli children's TV characters is Kishkashta, a prickly pear
cactus-shaped puppet (see Figures 2.2a and 2.2b).[7]

The idea of who or what is the Israeli Tsabar was also encapsulated in the
caricature image of 'Srulik', which has come to symbolise the nation and its
character. Interestingly, one of the first drawings of 'Srulik' was 'as the middle
portion of a prickly pear cactus plant, without thorns on his body but with ...
[the] intention of conflating the new native-born Israeli with the prickly-pear as
tsabar' (Bardenstein, 1998: 12).

There is some confusion over where and when the term tsabar arose to describe
Israeli-born Jews. Oz Almog (2000) and others have suggested it originated as a
derogatory term used by Jewish immigrants to Palestine in the 1930s to describe
the native-born Jewish population; tsabar being a Yiddish pronunciation of the
Arabic word for the prickly pear, *ṣabr* – the Eastern European 'ts' sound (the letter
'*tsadik*' in Hebrew) replacing the Arabic-Semitic 'ṣ' sound (the letter '*al-ṣād*' in
Arabic). For those European Jewish immigrants, native Jewish settlers might have
appeared wild, exotic and 'indigenous'; many of the native Jewish settlers indeed
adopted local customs, food practices, fashion and vocabulary.[8] From the late
1930s onwards, the term became synonymous with Jews born in Palestine, and
even a source of contention regarding who was a proper tsabar and what exactly

---

6   See for example, Meron Benvenisti [in Hebrew], *The Dream of the White Sabra*
(Jerusalem: Keter Books, 2012).

7   Kishkashta was the host of the children's TV program *How Come?* (the program
ran from 1976 to 1981). According to the Israeli Educational Television, he 'is a cactus-
shaped hand puppet who loves to play with words, songs, riddles and jokes' and his image
is related to the fact that 'Israeli-born children are known as "Sabras" – cactus plants that
are hard on the outside and soft and sweet on the inside'. The character of Kishkashta was
brought back in 2009 by the Israeli Educational Television in a new program titled *Who
Comes to Kishkashta?*

8   More on this in subsequent chapters.

Figure 2.2a   Kishkashta – the Israeli children's TV character; original black and white image 1976–1981

Source: Courtesy of the Israel Educational Television.

Figure 2.2b   Kishkashta from *Who Comes to Kishkashta* (2009–)

Source: Courtesy of the Israel Educational Television.

was tsabar culture and mentality. It has also come to embody, for many decades, what it meant to be an Israeli or a Jew in Palestine. The image of the tsabar and its culture were later epitomised by groups such as the Palmach, the elite combat unit of the Haganah, which was the main paramilitary force of the Zionist movement at the time. Members of the Palmach, who were mostly of European origin, though many of them were born and raised in Palestine, romanticised and exoticised the Arab-Palestinian *fallah* (peasant), with whom they wanted to share the sense of the significance of agriculture and the connection to the land. Nonetheless, and despite their desire to acquire Arabic language skills, they also wanted (just like members of Ha-Shomer and Ha-Roʿeh) not only to resemble the romanticised notion of the Arab-Palestinian 'other', but also to be a perfect substitute for him.

The story of the *şabr/tsabar*/prickly pear is an interesting and illuminating one. Strikingly, the cactus and its fruit originated in Central America and were brought over to Europe and North Africa by the Spanish, and therefore are not a 'natural' symbol of the land of Palestine/Israel. Yet, since it arrived with the Spanish, it has spread and flourished in the region due to the favourable climate. Visitors to Palestine in the first decade of the twentieth century described how ubiquitous it was; the cactus and its fruit were seen literally everywhere (see, for example, Grant, 1907; Masterman, 1901). According to Tal Ben Tzvi, its prominence as a major part of the Palestinian landscape was evident in local Arab-Palestinian paintings, songs and writings. For example, the Palestinian artist Zulfa al-Saʿdi presented an exhibition of oil paintings at the first pan-Arab fair in 1933 in Jerusalem juxtaposing heroic Arab historical figures, such as Saladin, with the image of the prickly pear. The image of the fruit, according to Kamal Boullata (2001), was already established by that time as a metaphor for village life and patience. Linguistically, the Arabic root of the word *şabr* is also related to the ideas of patience, endurance and resistance (an important aspect of Palestinian life in exile and under Israeli occupation). Nasser Abufarha (2008) explains that the fruit also played an important part in village life in Palestine: as mentioned before, the cacti served first and foremost as hedges that signified village boundaries. Additionally, the eating of the fruit, because of the difficulty in dealing with its many small thorns, became a village event that strengthened communal social bonds. The soft delicate and sweet flesh of the fruit contrasted against its thorny exterior and the harsh environment, provided a metaphor for village life – the tedious and hard work that went together with the strong social bonds and the beauty of the landscape.

This is not to argue that the prickly pear was a prominent Palestinian national symbol in the period leading up to 1948, or that its meaning remained constant, but that it was an ubiquitous part of local Arab-Palestinian culture, landscape and life. It would be hard to argue that Zionist setters were unaware of the Arab-Palestinian connection to the prickly pear when they started using the term; in fact, they might have started to use the term because of this knowledge. What is of interest is that this relationship between the indigenous Arab-Palestinians and the prickly pear has been mostly absent from Zionist writing on the origin and meaning of the tsabar. This denial of the Arab-Palestinian root of an Israeli national symbol was so profound that in the 1980s, when prominent Palestinian artist ʿĀṣim Abu Shaqra presented an exhibition around the theme of potted prickly pear cacti, some Israeli art critics thought this was a Palestinian way of engaging with the identity of the other, not recognising the interplay of symbols and meanings (Mohammed, 2006). However, the connection between the prickly pear and Palestinian nationalism became more noticeable in Israel with the outbreak of the First Intifada – the first popular uprising against Israel's occupation in the West Bank and the Gaza Strip – when it was reclaimed by Arab-Palestinians and used as a symbol of their steadfastness in relation to Israeli occupation.

One might argue that it is one thing to appropriate symbols that are connected with the *other* – the prickly pear, at least initially, was not overtly a national Arab-Palestinian symbol – but it is another thing to adapt and appropriate the *other's* national symbols. Today, and since the creation of the state of Israel, the Jaffa orange has been synonymous with Israeli agricultural prowess and Zionist achievements and success: the ability to make the desert bloom. For many years, the oranges have been one of Israel's main exports, and a leading brand recognised across the world. In recent years, Israel has even started to lease this brand to other states (Dror, 2004). As such, it is not surprising that the image was used in the Eldan car advertisement as a symbol for the state and nation. The importance of the fruit and the colour to Israeli national identity and the state has also been evident in Israeli art and design (Bardenstein, 1998),[9] as well as its educational policies. From kindergarten onwards, young Israelis learn about the importance of the orange, its uses and the role played by 'Jaffa oranges in the Zionist settlement of pre-state Palestine', alongside visits to orange orchards and orange-packing factories (Golden, 2005: 187).

In a similar fashion to the prickly pear, the story of the Jaffa orange also goes back to a period before the rise of Zionism. The mid-nineteenth century, particularly after the Crimean War, saw a rapid expansion of citrus production in Palestine. This was mainly due to the introduction of a new orange variety, the *shamouti* – whose name is based on the Arabic word for oval oil lamps (*shamout*) that were common in Jaffa at the time (Karlinsky, 2005: 16). The *shamouti* variety, because of its characteristics, tougher skin, strong resistance to bruising and drying up, lent itself more easily to exports (Shafir, 1996). In 1882, the year in which the first Zionist settlers arrived in Palestine, the Jaffa orange was already a known brand and made up about a fifth of the total exports of Palestine (Schölch, 1981). The cultivation, production and exportation of the Jaffa orange continued to expand rapidly in the late nineteenth and early twentieth centuries and, by the eve of the First World War, it had become the most important Palestinian export

---

9　See, for example, Ran Morin's well known outdoor sculpture '*The Floating Orange Tree*'. The sculpture depicting an orange tree sprouting out of a floating orange (an orange suspended in the air with the aid of wires), is located in the old city of Jaffa.

(Khalidi, 2010). The orange orchards, many of which were in the coastal region and in and around Jaffa, resulted in Jaffa being referred to as 'the city of oranges'.

Because of its importance to the Palestinian economy, the Jaffa orange was used discursively as a national symbol connecting Arab-Palestinians to the land of Palestine (Bardenstein, 1998), and the importance of the orange as a national symbol grew even more prominent after 1948 (see, for example Bardenstein, 2007; LeVine, 2005: 147). The symbol of the orange was also discussed in relation to a Palestinian flag in the late 1920s. Among the ideas generated for the design of a Palestinian national flag was the inclusion of the orange and/or the colour orange. The idea signified a desire among some Arab-Palestinians to stress their local Palestinian identity alongside their belonging to an Arab nation (Sorek, 2004).

The transformation of the Jaffa orange from a Palestinian into an Israeli national symbol was gradual.[10] The interest of the Zionist movement and of individual Zionist settlers in the citrus industry was stoked by the popularity and profitability of the Jaffa orange trade. As pointed out by LeVine (2005), Jaffa was also, until the construction of the Haifa port in the 1930s, the main port of disembarkation for new Zionist settlers, who were exposed to the orange orchards upon their arrival. The first Zionist orange enterprises were mostly based on buying Arab-Palestinian orchards and were limited in scope, but the movement towards Jewish-Zionist production was unmistakable (Karlinski, 2000). It is important to mention, though, that even during this period of limited Zionist orange production, concerns were raised by Arab-Palestinian merchants over the potential threat from Jewish orange exports, and in particular the use of the Hebrew language on the orange crates (Kark, 1990). Through a mixture of private Jewish, combined Jewish-Arab and Zionist official enterprises, the orange industry became an important sector in the Zionist settler economy and its leading export. The inflow of foreign investment, technological innovations and the constant influx of new settlers was vital to the Zionist orange industry, which expanded dramatically in the 1920s. By the late 1930s, the Zionist orange industry was responsible for around half of all orange exports in Palestine (Karlinski, 2000). Throughout the British Mandate period (1919–1948) and despite the rapid expansion, industrialisation and modernisation of the Zionist settler economy, the Jaffa orange remained Palestine's main export.

In the aftermath of the 1948 war, and as a result of its popularity, the Jaffa orange brand and symbols were retained and used by the state of Israel.[11] This occurred despite the continued use of the symbol by Arab-Palestinians in Palestine

---

10 In this regard, a movie made by Israeli documentary film maker Eyal Sivan contributes interesting insights on the transformation of the Jaffa orange from a local Arab-Palestinian fruit and symbol of production into an Israeli export labelled '*Jaffa* oranges'. Sivan's movie, made in 2009, is titled: *Jaffa: The Orange's Clockwork.*

11 One could argue instrumentally that there was little economic sense in not retaining this lucrative trademark.

and in the diaspora – after the war, the symbol of the Jaffa orange became closely associated with the Palestinian *Nakba* and the loss of Palestine.[12] However, and as noted by Raviv (2003: 21), through its appropriation by the state of Israel, the Jaffa orange was 'estranged' from its Arab-Palestinian history, 'while presenting Palestine as empty and desolate before their arrival'. In Jewish-Israeli mythology the Jaffa orange became a symbol of Zionist pioneering spirit and or socialist Zionist achievement of making the desert bloom.

National Flora and Fauna of Israel/Palestine

Following the orange and the prickly pear, the country's flora and fauna, and the Israeli national symbols within this field, serve as an interesting case study. This is because nature is not a concept that was 'made' by the local Arab-Palestinians, but that was conceptualised as being always 'there'. The imagined permanence of the landscape, flora and fauna – also made it easier to link to the Bible and thus to link the new Jewish nation to its perceived and/or imagined roots, history and heritage.

In this regard, and following Dahl (1998), we found that '"nature loving" as nationalism' was and still is an important element that has helped to bolster the Jewish Zionist connection to the land, and as a result has played an important part in the emergence of the natural 'national symbols' of Israel. 'Nature loving', in our case, was particularly exemplified in the Zionist educational policy and pedagogical principal of '*yedi'at haarets*' (lit. 'knowing the land'). The central purpose of this idea, which was also a school subject, was to instil Jewish national and Zionist values and ideas in the Jewish population through the act of travelling and hiking in the land of Palestine/Israel while learning about its history and geography (Benvenisti, 2012). Learning about the land, and its nature and geographic features thus became intertwined with Jewish-Israeli identity and national symbols.

A good point of departure for symbols in this category of nature, flora and fauna, is 'the native dog of Israel': the Canaan dog. This breed has been part of the region for thousands of years, and one of the first earliest indication is found in the ancient cemetery of Beni Hasan in Middle Egypt, where drawings on tombs that go back to 2200 BC depict dogs that look similar to the Canaan breed (Palika, 2007: 184). As Israeli dog expert Eyal Murad (2006) highlights, the dog has biblical as well as modern connections. According to him, 'dating back to Biblical times, the Canaan dog enjoys the distinction of being the world's oldest new dog breed. ... The national dog of Israel, the Canaan dog's ancestors once served as herding dogs for the Israelites' flocks'. This short depiction is an interesting one as it sheds light on a few important elements. It underlines the

---

12   See, for example, Ghasan Kanafani's classic short story 'The Land of the Sad Oranges' about the Palestinian Nakba following the 1948 calamity.

biblical importance of the Canaan dog; it implies the geographic significance of 'Land of Canaan' – the territory on which modern Israel is located; it ties together the past and present ('the oldest new dog breed'); and it stresses the dog's biblical expertise as being a good 'guardian' and 'herder' – exactly the same characters for which it is still famous in the twenty-first century.

One should ask, however, what has helped to push forward this particular breed as the national dog of Israel? Interestingly, this has had to do with the indigenous Arab-Palestinian population more than anything else. As Palika (2007) highlights, the dogs were used by Bedouin communities in the region 'as both guard dogs and herding dogs'. Hence, in the encounter between the Jewish immigrants and the Bedouin, and in their attempts to push forward 'Hebrew labour' by creating a Jewish movement for guarding (Ha-Shomer) and a Jewish initiative for shepherding (Ha-Ro'eh), the supporting Bedouins' dog – together with the skills of herding and guarding – had become a pioneering Zionist-oriented breed. At that moment in time, the dog was not called 'Canaan', but was simply known as an 'Arab dog'. In the memoirs of the first Jewish watchmen (lit. *Agudat Ha-Shomrim*) one of them remembers that 'we had two Arab dogs with us (from Canaanite breed) to which I brought meat and bones' (Palika, 2007: 379).

This initial use of the dogs by the Jewish watchmen was later taken forward into even more explicit paramilitary services. When Dr Rudolphina Menzel, a Vienna-born cynologist, arrived in Palestine in 1934, she was asked by the leadership of the Haganah to find a breed which would be suitable for protecting the newly built Jewish settlements and to enable the continuation of the Zionist project in the country. Menzel came up with the idea of finding scavenger Canaanite/Arab dogs in the desert area (the Negev/*al-Naqab*) as well as trained dogs from the Bedouins' to be used as guard dogs in the new Jewish settlements, especially those that bordered Arab villages. As Kaftal and Boyd (2000) highlight, following her decision, Menzel initiated a selective breeding programme to reproduce this breed, and eventually the dogs were used in different security missions – including during the 1948 war. Following the creation of Israel, Menzel created a habitat for the breeding of the Canaanite dog, in the *Bāb al-Wādī/Sha'ar Ha-Guy* area and, following the dogs' defence orientation, called it 'sons of security' (lit. *bney ha-biṭaḥon*). In 1966, Professor Menzel was honoured as the first person in the world to define the breed-standard of the Canaanite dog, and for ensuring that this breed – despite its Bedouin connection – would be called the 'Israeli, Canaanite dog'. In a similar way that Jaffa oranges were turned into Israeli Jaffa oranges, the Arab-Bedouin dog was transformed into Israel's national breed of 'Canaanite dog'. Interestingly, Myrna Shiboleth – a world champion dog breeder – calls the Arab/Canaanite dog, 'A real Sabra dog'. Furthermore, in an interview with *Haaretz*, Shiboleth further erases the Arab-Bedouin-Palestinian connection to the dog, as she states that 'All the Canaan dogs in the world are descendants of Israeli dogs' (Lior, 2012). This completes the 'life cycle' of this specific species in the last hundred years: the 'Arab-Bedouin dog' was turned into a

'Canaanite dog' and, following a process of being appropriated by the 'Hebrew man' and related to the Jewish Bible, was in turn made into an *Israeli dog*.

Staying with the world of nature, and its national appropriation, the de-Arabisation and Israelisation processes seem even clearer when analysing the former national flower of Israel: the *cyclamen persicum* (lit. *rakefet*). The white, pink, purple or reddish-coloured flower, which grows naturally from October until April, especially in the mountainous regions of central and northern Israel, has become a national symbol of the state of Israel. The flower was represented as Israel's 'national flower' on several occasions (for example, in 2007, at the international flower competition in Beijing). One of the reasons the flower had been elevated into a national symbol was that it grows underneath rocks in the mountains, which combines characters of steadfastness, beauty, shyness, and localness, and in one way or another recalls the set of values 'desired' by the Zionist tsabras. In the Jewish-Israeli culture, the flower is also frequently anthropomorphised as being 'shy' and 'humble'.[13]

The flower has also had a special place in Jewish popular culture over the last hundred years, best exemplified by its mention in the 1921 song '*Rakefet, rakefet*', composed by Levis Kipnis in Tel Hai in the upper Galilee, just a year after the battle of Tel Hai, a battle seen as symbolic of Zionist heroism. The song, which was adapted into Hebrew from the original Yiddish, symbolised the dramatic moment of the alleged Jewish transformation from the diasporic victim Jew to the local, brave, Hebrew man. Tel Hai is also forever ingrained in Zionist and Jewish-Israeli memory and mythology through the final words attributed to Josef Trumpeldor – a Zionist military leader who died as a consequence of the battle there – 'it is good to die for our country'. The song, therefore, written a year after the battle in Tel Hai, came to be associated with the Jewish return to nature, and the return to Palestine/Israel.[14]

Yet, despite the 'Jewishness' or the 'Israeliness' of the rakefet, there are elements of the flower that have been ignored over the years. First, the name rakefet – despite sounding Hebraic, does not originate from Hebrew and, while it is mentioned in Jewish legends, it is never specifically mentioned in the Bible itself. The name 'rakefet' derives from the Aramaic *rakafta*, which in turn was probably the root of the flower's Arabic name: *rakaf*. This was the name to which

---

13   For further reading on the *rakefet* in Israeli culture, see: The Official Website of Wild Flowers of Israel: http://www.wildflowers.co.il/hebrew/plant.asp?ID=61 (accessed: 1 February 2015). See also: Menashe Geffen [in Hebrew], *Transformation of Motifs in Folklore and Literature* (Jerusalem: R. Mass, 1991), p. 45.

14   For further reading on the importance of Tel Hai see: Tamar Mayar, 'National Symbols in Jewish Israel: Representation and Collective Memory', in Michael E. Geislar (ed.) *National Symbols, Fractured Identities: Contesting the National Narrative* (Lebanon, NH: University Press of New England); and Yael Zerubavel (1991) 'The Politics of Interpretation: Tel Hai in Israel's Collective Memory', *Association of Jewish Studies Review*, 16(1/2): 130–60.

the first Zionists were introduced when they encountered the flower while getting to know the country and its nature from their Arab-Palestinian guides, shepherds or watchmen.

In addition, the flower – despite being very central to the Israeli culture – has only one name in modern Hebrew (which is based on the Arabic/Aramaic name), while in Arabic it has 12 additional names. This suggests a profound attachment between the raḵefet and the local Arab culture. Dafni and Khatib (2001) argued that perhaps of all the flowers in the country, the raḵefet is the 'champion of names' in Arabic, and 'there is no other local wild plant that has as many names in Arabic as this one', a notion that was mentioned also by Atatreh (2014) and Shirkis (2008), with a focus on the Arab-Palestinian popularity of the flower.

The name *rakaf*, means 'light snow' in classical Arabic, probably referring to the flower's white colour and the fact that it 'hides' underneath rocks, where an accumulation of frost, slush and snow first appears in the winter. However, the flower also has additional names; several of these relate to the flower's shape, colours and the area in which it grows,[15] other names relate directly to Arab and Arab-Palestinian history, folk and traditions. For example, the flower is also known as *kharūf*, meaning 'sheep'. According to local Arab legend, around the Carmel mountain, where the flower grows, children used to take the flowers of the raḵefet/rakaf and string them into one another (the stalk into the flower) in order to create a white necklace that looked like the mane of a sheep. The flower is also known as *ṣabūn al-rā ʿī* (lit. 'the soap of the shepherds'). This name hints at an old Arab use of the flower. Apparently, the bulb of the flower contains foaming ingredients (saponins) that help to remove oily stains from cloth. The Arab shepherds used to rub the oily stain with a halved bulb, and then wash the cloth with water. Interestingly, the English name of the flower, *cyclamen*, is based on the name of a popular saponin. Another name, *tāj Suleiman* (lit. 'the crown of King Solomon') is based on a legend that the shape of the flower inspired the creation of the crown of King Solomon. Interestingly, the flower is also related to Christianity. One of its names, *bakhūr Maryam* (lit. 'the incense of Virgin Mary'), hints at an Arab-Christian origin, and connects the flower's fragrance with the smell of the church's incense. The flower's famous Arabic 'culinary name' is simply *ṭuʿm* ('tasty') or *za ʿmaṭūṭ* (unknown etymology), referred to when the flower is used in the kitchen, usually to make wrapped leaves stuffed with rice and meat. Another use of the flower in the local Arab culture, this time

---

15 For example, the flower is referred to at times as *raqaf*, which means in Arabic 'slight movement'; *shqūq*, meaning cracks or fissures; *ḥamīra sqāqa*, which means the 'red one in the cracks'; *udhn al-arnab*, meaning 'the rabbits' ears'; *Qarn al-ghazāl*, meaning the horns of the gazelle; *ṣurm al-dīk*, meaning 'the anus of the chicken', or *ṣrīmat al-jaj*, meaning 'the little anus of the hens'; *dwīk al-jabal*, meaning the 'small chicken of the mountain'; *juz al-ḥamām*, meaning 'a pair of lovebirds'; *ʿaṣat al- rā ʿī*, meaning 'the shepherd's walking stick'; and *ghalyūn al-qāḍī/ghalyūn al-ḥākim/ghalyūn al- rā ʿī*, meaning 'the pipe of the judge'/'the pipe of the wise man'/and 'the pipe of the shepherd'.

not by shepherds or cooks, but by fishermen, is *munawwim al-ḥūt/munawwim al-samak* (lit. 'the anaesthetiser of the whale'/'the anaesthetiser of the fish'). This recalls an old fishing method of creating small balls of the flower's crushed bulb with flour, which were spread on the water. Fish which ate this bait were 'drugged' and floated and were then collected by the fishermen, who had to clean the fish's guts of the poisonous bait before they could eat them.

The richness of names allocated to this flower demonstrates its profound cultural role in the region. It was mentioned in connection to historical religious figures (from the Virgin Mary to King Solomon); in relation to the local Arab shepherding culture (which was very much desired by the first waves of Zionist immigration to the country); in relation to folk legends and aesthetics; and professionally as part of the local cooking and fishing culture. Perhaps this central role of the flower within indigenous Arab-Palestinian culture, paved the way for the raḳefet to become truly 'Israeli'. The centrality of the flower to Arab culture hinted both at the uniqueness of the flower in Palestine in comparison to the wider region (a 'flora and fauna' uniqueness which was prized by the first Zionists) and to its role as a signifier of a *local* culture. As was seen before, it was almost entirely through the appropriation and transformation of *local* symbols that were then de-Arabised that they were seen as local and unique by Jewish-Israeli society, and as 'Erets Yisraeli' and 'Israeli' in their very nature.

Another flower, which is perhaps not as celebrated as the raḳefet, but that is usually in competition for the status of 'the flower of Israel', is the *kalanit*. This flower, known in English as anemone coronaria or the poppy anemone, is known for its beauty and this is reflected in the etymology of the name in Hebrew, which indicates the connection between the flower's name and the word *kala*, which means 'bride' in Hebrew.

The flower is called *shaqāiq al-Nuʿmān* in Arabic, which means 'the wounds of Nuʿmān', after the Mesopotamian god of food and vegetation, Tammūz or Nuʿmān in Arabic. According to the legend, when Nuʿmān was on his way to his beloved goddess of Ishtar, a boar attacked him, and wounded him badly. When he was carried by his friends to Ishtar, every drop of red blood that fell onto the ground later bloomed as a beautiful anemone coronaria, thus creating a carpet of beautiful red flowers. According to Galai, the name in English, anemone, is a distortion of the original Arabic, Nuʿmān (1988: 10) and one could argue that the connection between blood and flowers is also replicated in other communities' ceremonies surrounding commemoration.[16]

In a similar way to the cyclamen, it is interesting to note the way in which the kalanit made its way from a symbol of Arab-Palestinian nature to one of Jewish-Israeli nature. For example, the 'return' to *Erets Yisraeli* symbolised also the return to its nature, as can be shown in the case of the kalanit. The flower is used

---

16 In the UK, the annual Memorial Day for the fallen soldiers is marked by the poppy flowers. Interestingly, the poppy flower, due to its similarity to the anemone coronaria, is sometimes called poppy anemone.

in the coat-of-arms of several Israeli cities and towns – Karmiel in the Galilee most famously; a town which has come to be seen as part of the project of de-Arabising and 'Judaising the Galillee'. Moreover, the late Prime Minister Ariel Sharon was buried outside his farm in the south of Israel, on a hill that is now called 'Anemone Coronarias Hill' (lit. Givʿat ha-Kalaniyot), and like the famous song of '*Raḵefet, raḵefet*', also the song '*Kalaniyot*' has become one of the most famous Israeli songs of all time. The song, written by celebrated Jewish-Israeli songwriter Nathan Alterman in 1945, quickly came to represent the 'return' to the land, following up on the fact that the flower totally disappears in the summer but *returns* every winter. Another explanation for the song was Alterman's sadness in light of the horrors of the Second World War, and his desire to see some hope, in which flowers will return (Laor, 2003).

The 'tension' between the alleged *Israeliness* of the flower on the one hand, and its importance and symbolism within Arab-Palestinian culture is obvious. On the one hand, its 'Israeliness' is proven in its symbolic relation to the Second World War, to the act of the Jewish *return* to Erets Yisraeli, as well as to the Hebrew word *kala* (bride), to the burial place of former Prime Minister Ariel Sharon, and as the symbol of several Jewish cities and towns. On the other hand, the Arab relation to the *shaqāiq al-Nuʿmān*, the same flower but in Arabic, is rooted in the flower's connection to Arab folklore and culture, as well as its many usages in medicine.

The following example about the use of the two flowers mentioned above in both Jewish-Israeli and Palestinian societies seems to capture this tension well. In 2007, as part of the many international events that were held in Beijing in the lead-up to the Olympic Games, an international agricultural exhibition was held. Israel sent three items as its national symbols: an olive branch – a symbol of peace, a very dominant tree in Palestine/Israel, a symbol of the Palestinian struggle for independence, and one of the most famous products of Palestinian farmers: olive oil. The second item that Israel sent to the competition was a bouquet of anemone coronarias. The third item was a bouquet of cyclamen. The Arab press could not contain its fury over what they reported as the 'total appropriation of Palestinian legacy by the Israeli occupation'. Al-Jazeera's headline at the time was 'The Israeli Occupation Chooses *Shaqāiq al-Nuʿmān* and *Qarn al-Ghazāl* to Represent It Abroad', highlighting that the 'Israeli state of occupation chose Palestinian flowers as Israeli symbols. … They chose two flowers that have such a strong connection to the Arab culture, to Arab literature and folklore, to Islam, and to the Arab kitchen' (al-Rajoub, 2007).

Following this, the Palestinian News Network, posted a news item saying that 'The Fight Is Over Symbols', and in the article it gave a list of Palestinian officials who responded with anger to the Israeli delegation. They said that it is unbearable that these flowers will be put next to an Israeli flag, mentioning their connection to the Palestinian culture, such as 'the colours of *shaqāiq al-Nuʿmān*, which resemble the colours of the Palestinian flag, and have become a symbol of the blood of the Palestinian martyrs'. Imad al-Atrash, from the Palestine Wildlife Society, said that 'after Israel stole the land, they began stealing its flowers as well'. A representative of the Palestinian Ministry of Agriculture, said it is 'a disgrace that China did not send an invitation to the Palestinians, and instead they ended up having an Israeli team that uses Palestinian occupied lands as theirs', highlighting that by so doing Israel 'erases the Palestinian legacy in Palestine'.[17][18]

In China, therefore, the tension over Israel/Palestine seemed to be clearer than ever. As a matter of fact, any Israeli item that was 'from the land' would be straightforwardly considered by Palestinians as theirs, while for Jewish-Israelis it is obviously Israeli as it comes from the land that they left and to which have now 'returned'. The struggle over the indigenous symbols of Palestine/Israel encapsulates the fight of *indigenousness* and *indigeneity*, which are arguably at the very heart of the Israeli-Palestinian conflict, as the case study of these symbols demonstrates. This tension lies also in the mere question of who the anemone coronarias belongs to. For the Arab-Palestinians it is a local flower which symbolises their land and heritage and, for Jewish-Israelis, it is a local flower that symbolises the land to which they 'returned'. It seems that this is also the tension that separates 'borrowing' from 'taking back', 'adaptation' from 'appropriation', which is as crucial in the case of the flowers as it is in the case of the land itself.

This tension became even more apparent to us the day before we submitted our book to the publisher, when we received an email from a few colleagues about a new social media uproar in Israel regarding gazelles. The story of the gazelle, the Arab symbol of beauty (*ghazāl*) being turned into an Israeli national symbol is another example of how the land's fauna was detached from its Arab origins through a process of Hebraisation and biblicalisation.[19] It also points to the continuing struggle in Israel over symbols, ownership and indigenousness.

The social media uproar began when a popular far-right Jewish-Israeli hip-hop singer, known as 'the Shadow', called attention to the fact that, at the Jerusalem Biblical Zoo, the name given to the native gazelle was 'Erets Yisraeli gazelle' (lit. '*tzvi Erets-Yisraeli*') in Hebrew but 'the Palestine gazelle' in English and Arabic. 'The Shadow' asked in his Facebook post 'a question to all those garbage-can geniuses on the left – why is it called Israeli in Hebrew and Palestinian in English? What do you say to that, arrogant, self-righteous and insufferable scum?'

---

17   For further reading, see the Palestine News Network's article in the Israeli website kibush (the original Arabic article was erased from the website): http://www.kibush.co.il/show_file.asp?num=22745

18   We have been unable to find any official Israeli response to this claim. However, it is clear that the Arab origins and associations with the flower, as well as with the raḳefet, are barely mentioned in Israeli media and literature, which view the flowers solely as Jewish-Israeli symbols. One of the few exceptions to this is the official website for the Society for the Protection of Nature in Israel. The society does acknowledge and refer to the Arab association of the flowers. See, for the anemone coronaria: http://www.teva.org.il/?CategoryID=200&ArticleID=21848, and for the cyclamen persicum: http://www.teva.org.il/?CategoryID=200&ArticleID=612.

19   That is, linking a story or item to a number of biblical passages that proved its symbolic connection to modern Jewish national identity.

The Biblical Zoo's response was that the gazelle has a Hebrew name (that is, unsurprisingly, based on and linked to a number of biblical passages and stories) as well as a scientific name, the 'Palestine gazelle'. The name 'Palestine gazelle' was given to this subspecies of the mountain gazelle, a species of gazelle that is native to the Middle East, because it is native to Palestine (Yaron, 2015).[20]

However, and as exemplified by this story, there is an inherent Jewish-Israeli fear that any mention of the word Palestine and/or a connection between the Arab-Palestinian people and culture and the land, including its flora and fauna, is an attempt to question the modern Jewish link to Erets Yisraeli. This is despite the fact that the adoption and appropriation of the local flora and fauna has been based on turning these symbols from geographic, regional, intercultural and borders-free representations into Jewish national symbols, of *one* people, and of *one* – blue and white – nation.

Everyday Symbols

The way in which an Arab, or an Arab-Palestinian element was transformed into a Jewish-Israeli, pioneering, 'original', symbol, with or without a connection to a real or imagined Jewish biblical past, can also be seen in some of the symbols that have contributed to the 'mundane' Jewish-Israeli spirit and culture. As we will demonstrate, this appropriation of mundane symbols, away from official and/ or elite control, shows that the process of creating the national symbols of Israel was not simply a top-down one, starting from the state and its institutions and permeating to its people, but often the opposite, a bottom-up process percolating upwards through society.

This process often began from the 'grassroots' level, from the social and political encounters between the early Jewish-European Zionist immigrants to the country and the Arab-Palestinian people. This encounter set in motion the process of forming and crystallising the Jewish-Hebrew-Zionist culture, which was based on local (Arab) elements on the one hand, and a gradual, larger process of denying this connection, on the other hand.

The story of the Israeli sandal is probably a good starting point for our analysis. In the edited work *Jews and Shoes*, Ben-Meir (2008) highlights how what is today considered 'the Israeli sandal' went through a process of 'invented tradition', whose roots are found in Zionist ideology and in the period of the Second *'aliya*. Ben-Meir traces a very interesting process. On the one hand, for members of the Second *'aliya*, these sandals, which were perceived to have been invented in the kibbutz, served as a means of negating the diaspora (sandals and sunny Palestine in contrast to boots or shoes in cold and snowy Eastern Europe) as well as a symbol of the new agriculture-oriented, modest, democratic and gender-neutral, pioneering

---

20   By and large, flora and fauna with 'Palestine' (or Palaestina) as part of their scientific name have been translated into Hebrew as 'Erets Yisraeli'.

Jewish person, the tiller of the soil (Raz, 2002). On the other hand, she found out that while the sandals were, first, associated with the kibbutz, at the end of the 1930s and beginning of 1940s, and more popularly in the 1950s and 1960s, their origin was transformed and reinterpreted as Jewish and they became known and labelled as 'biblical sandals' (lit. *sandalim tanakhiyyim*). The transformation of the sandal into becoming an Israeli, biblical, Hebrew, and Sabra symbol occurred gradually and at different levels, and represented the fulfilment of various national needs as encapsulated in the religious-secular, past-and-present reasoning of the Zionist movement. In this light, the sandal was important as it was the shoe that one puts on the ground and so symbolises the physical 'redemption' and control over the land. The sandal had an added importance as it highlighted the connection between the modern land and the 'nation's past' – real or imagined – as well as the connection to the nation's biblical 'ancestors'. Combining these two factors together, one can better understand how the sandals were related to central Zionist ideas of 'reclaiming' and 'redeeming' the land: tilling the land (representing the principle of 'conquest of the land' – lit. *kibbush ha-adamah*); replacing the Arab-Palestinian *fallah* with a Jewish farmer (representing the principle of 'conquest of labour' – lit. *kibbush ha-ʿavodah*); and creating Jewish-Hebrew pioneers in the fields of shepherding, guarding, farming and hiking – in line with Ben-Gurion's concept of the 'conquering pioneers' (Segev, 1999: 209).

These sandals, therefore, whether secular and pioneering 'kibbutzinks' or 'ancestor-oriented' biblical sandals, have become a Jewish-Israeli icon. The famous animated figure of the Israeli person, named Srulik (a possible equivalent – at least from the symbolic point of view – to the Palestinian *ḥanẓalah*, who – unlike Srulik – is always barefoot) obviously wore these sandals. The website Israel-Catalog.com, for example, described the sandals as 'trendy Biblical-style sandals. … Feel the soft leather lining and flexible sole, the Biblical experience at its best' (quoted in Silverman, 2013: 20). Today, there are Israeli companies – such as Nimrod, Shoresh and Teva – that specialise in making these sandals and that have become famous worldwide for this 'Israeli product'. Most interestingly, in contemporary Jewish-Israeli society they are no longer as popular among secular types as they are among the national-religious groups, and especially the settlers' movement (Pedhazur, 2012: 113; Peleg, 1997: 60), a point to which we will come back to in the fourth chapter.

Strikingly, while the sandal has been connected to the Jewish biblical past (e.g. 'Bavta's sandal' in the Israeli national *Shrine of the Book*),[21] marketed by Israeli companies and appropriated by the settler-movement, its linkages to the Arab-Palestinian 'other' have been forgotten and marginalised. From the similar etymological root – sandal in Hebrew and ṣandal in Arabic – and up to the very centrality of sandals in Arab cultures, especially of desert dwellers like

---

21   For further reading on the style of the sandal and its influence on footwear in the Zionist movement, see: Tamar El Or (2012) 'The Soul of the Biblical Sandal: On Anthropology and Style', *American Anthropologist*, 114(3): 434.

the Bedouin community, the fact that the 'Israeli' and 'biblical' sandal is also Middle Eastern, Arab, Bedouin and Palestinian, has almost disappeared from the discourse in Israel.

Sandlas, obviously, are neither a 'Jewish' nor a 'Zionist' clothing invention. They are a product of living in hot desert conditions worldwide, and in particular in the Middle East. In *The Legacy of Tutankhamun Art and History*, Zaki mentions that sandals worn by ancient Egyptians 'comprised a sole, a thin strap, which passed between the big and second toes, two fastenings, which passed under the instep and sometimes a third, enclosing the back of the foot to hold it securely' (Zaki, 2008: 130). Also in *Life in Ancient Mesopotamia*, Mehta-Jones highlights that '[M]ost Mesopotamians wore leather-soled sandals with heel guards to protect their heel' and that 'until the time of the Assyrians, even soldiers wore sandals into into battle' (Mehta-Jones, 2005: 14). These descriptions, from Iraq and Egypt, which are suspiciously similar to the 'biblical sandals' made in Jewish-Israeli factories, are simply indications of the fact that sandals were neither Jewish, nor Christian or Muslim, they were just a very practical form of footwear in the region. Therefore, it is most likely that the kibbutznik's 'invention' was neither related to the Bible nor to a very original out-of-the-box sense of fashion. Most Jewish-Zionists in Palestine, it is evident, at the end of the nineteenth century and beginning of the twentieth century, who wanted to become more 'local' in the country romanticised and imitated two Arab 'types', the Bedouin and the peasants (the *fallah*s) – who, at times, wore sandals.

Bearing in mind that the sandals were used by desert dwellers, from the High Desert in Oregon 10,000 years ago to the Australian desert's indigenous population 5,000 years ago, to the Arab dwellers in the Nubian desert in the nineteenth century, it would not be a striking fact to reveal that the Bedouins too used sandals when herding their sheep in the Judean desert and in the Negev (Al-Naqab), which were made from 'camel's skin, and tied on with leathern thongs' (kitto, 1841: cclii). According to Rinat et al.: 'the Jewish settlers encountered their [first] sandals on their Arab neighbours' feet. In cold Europe this kind of shoe could not have developed, but in the country [Palestine] and in the Mediterranean climate, this [Arab] sandal has become a central summer footwear' (Rinat at al., 2009). This does not mean that biblical figures did not use sandals, but just that it is most likely that the inspiration for this 'pioneering', 'new Jew', 'biblical' footwear, which is today very much associated with the right-wing settlers' movement, originated from a direct encounter between Arab-Bedouin and Zionist immigrants, and not from the latter's reading of the Bible. In all likelihood, and in order to transform them into 'Jewish' footwear, and later a Jewish national symbol, the sandals were reinterpreted and contextualised as being part of a Jewish biblical tradition, 'preserved by the local population', to which the Zionist settlers returned.[22]

---

22   It is important to note that, while the early adoption of the sandal was inspired by the settlers' early encounter with the local Arab-Palestinian population, by the 1930s and 1940s Jewish, and later on, Israeli companies were producing their own sandal designs. For

The Kūfiyyā: Appropriating and Discarding Symbols

The story of how the *kūfiyyā* – the traditional Arab head dress – gained and then lost its status as an Israeli national symbol is a particularly interesting and illuminating one. According to Shapira (1992: 94) the first Zionist settlers often romanticised Bedouin culture and identity, including lifestyle practices, such as horse riding and the use of fire arms, 'power symbols' and dress, including the wearing of the kūfiyyā. In later periods, and well into the twentieth century, the symbolic importance of the kūfiyyā grew stronger. This went hand in hand with the crystallisation of the new Jewish image within the Sabra elite, and especially within the newly formed paramilitary organisations. In these organisations, in particular in Ha-Shomer and later in the Palmach, the adoption of the kūfiyyā was directly related to 'the flaunting of Orientalism in Sabra culture' evident in the adoption of Arab apparel and customs (Almog, 2000: 198). These, according to Almog, included the sandals and the kūfiyyā, as well as the drinking of 'black coffee', the 'crossed-legged sitting' (lit. *yeshiva Mizrahit* – 'Oriental sitting') and the adoption of Arabic words into Hebrew, elements that will be analysed in the following chapters. The Israeli writer Yoram Kaniuk, who was a member of the Palmach, remembered that:

> In the Palmach, but even before, we all tried to learn from the Arabs who were the original, the natives, not in culture but in manners: the kūfiyyā, the many Arabic curses, the coffee drinking … there was a need in our parents to see in us a new breed …. (In: Nocke, 2006: 204)

Ha-Shomer and the Palmach, as already mentioned, were important 'factories' in which Arabic items and customs became Hebrew and later Israeli-oriented. It was in these types of spaces and organisations that the customs of the Arab 'Other' were adapted into the body of the 'new Jew'. As we see it, it is not a coincidence that in these military-oriented spheres that aimed to contribute to the Jewish Zionist defence and 'common good', the use of the Arabic language, as well as the adoption of Arab customs and Arab dress, did not raise suspicion. Following Bartal's analysis of the fear and liberation paradigm, we believe that precisely in those military-oriented spheres there was no fear of becoming distant from and/or betraying the Jewish cause while being 'too close' to the Arabs. In one way or another, the importance of the military sphere foreshadowed that of the Military Intelligence sphere in Israel, as analysed by Shenhav, as the only one that 'permitted' Jews to use Arabic on a daily basis.[23]

---

more on sandals in Israeli society, see: Tamar El Or [in Hebrew], *Sandals: The Anthropology of Local Style* (Tel Aviv, Am Oved, 2014).

23   For further reading on Shenhav's notion of the 'security license' to speak Arabic in Jewish-Israeli society, see: Yehouda Shenhav, *The Arab Jews: A Postcolonial Reading of Nationalism, Religion, and Ethnicity* (Stanford, CA: Stanford University Press, 2006), p. 3.

This 'entrance' of the kūfiyyā into the Hebrew dress also uncovers the 'double relationships' that the first Zionists had with the local Arab-Bedouin 'other', which was at the same time imagined as the biblical Jewish 'self'. A famous photo of Chaim Weitzman, from 1918, (see Figure 2.3) shows him – the head of the Zionist Organisation and later the first President of the state of Israel – meeting with the Emir Faisal, representing the Arab Kingdom of Hejaz and Greater Syria (and later King of Iraq as well). In the meeting, the two agreed on principles for Jewish immigration to Palestine (following the Balfour Declaration) and Jewish–Arab relations in the future Arab state. The interesting element for us in that meeting, was the decision of Weitzman to show up wearing a kūfiyyā on his head. Weitzman, who was born in Pinsk in the Russian Empire, did so as a sign of respect for the Arab Emir, but also as sign of the closeness between the two people who were about to discuss their future together.

Even David Ben-Gurion, the first Prime Minister of the state of Israel, had an interesting 'self' and 'other' relationship with the kūfiyyā. A photo from the 1948 war is one of a number of telling examples of the appropriation of the kūfiyyā by Ben-Gurion, who was born in the Polish town of Płońsk. In the photo, Ben-Gurion is seen leading a group of soldiers in a battle. The interesting element is that Ben-Gurion is seen in the photo holding binoculars in his left hand, putting his right hand in his pocket, exposing the pistol placed on his belt and, most importantly for us, revealing the Bedouin kūfiyyā around his neck (see Figure 2.4). Ben-Gurion's use of the kūfiyyā was not limited to one single event. In fact, there were numerous occasions in which Ben-Gurion wore the kūfiyyā (see figures 2.5 and 2.6). Ben-Gurion, as we argue, was not wearing the kūfiyyā simply as a sign of respect for Arab-Palestinians. His images with the kūfiyyā, similar to its use by members of Ha-Shomer, the Palmach and Jewish farmers throughout the late Ottoman Period in Palestine and the British Mandate in Palestine, represented the Israeli brave 'new Jew', who now saw himself as the native, who knew the landscape of 'his' country and who travelled throughout it without fear.

During the 1950s and 1960s, for example, Ben-David identified the dominance of the kūfiyyā in the outfit of Israeli hikers and tour guides, mostly those associated with the Society for the Protection of Nature in Israel. Their unofficial 'uniform' was strikingly similar to that of the Palmach – which was a take on the Bedouin dress – and included either wide-brimmed hats or kūfiyyās (Ben-David, 1997: 132).

In discussing her book on important milestones in Israeli fashion, Nurit Bat-Yaar echoes the biblical justification for the use of the kūfiyyā, writing that 'Israel's early settlers regarded it [the kūfiyyā] as a local accessory which related to Middle Eastern cultural sources ... and during the Palmach period it became the coolest accessory around, donned even by Prime Minister David Ben-Gurion'. Bat Yaar also mentions that the combination of the early encounter with the kūfiyyā,

Figure 2.3   Israel's first President, Chaim Weitzman (at the centre) with the Emir Faisal

Source: Courtesy of the Central Zionist Archive.

together with later encounters,[24] increased the use of the kūfiyyā, or of similar shawls, in Israel. According to her, one of the leading Israeli fashion designers, Simi Lederman Elbaz, created a kūfiyyā:

> which became popular in the 1970s as part of Eilat's Club Mediterranée's boutique. ... So popular were her colourful kaftans, dipped in Eilat's shades of violet, turquoise, clear green, yellow, white, orange and red that when Paris couturiers André Courréges and Ted Lapidus visited Eilat's Club Mediterranée they didn't leave before tucking in their suitcases Simi's Kafiya Kaftans.[25]

24   For example, the Israeli encounter with the Bedouin tribes living in the Sinai Desert following its occupation by Israel during the 1967 war. As a result of Israeli-Egyptian peace negotiations, Israel gradually withdrew from the territory, and it was finally returned to Egypt in 1982.

25   For the original quote, see the official website of Nurit Bat Yaar, titled *Fashion Art*: http://nuritbatyaar-fashionart.blogspot.co.il/2010/08/israeli-fashion-kafiya-andre-courrege. html (accessed: 2 January 2015)

Figure 2.4   Israel's first Prime Minister, David Ben-Gurion (in the centre) with a group of soldiers

Yet unlike the previous symbols we discussed above, the transformation of the kūfiyyā from a Bedouin dress into an Israeli 'national' symbol was not completed. This dress item, especially in the 1970s and 1980s lost its 'local' scent and – perhaps also due to the worsening political situation, the 1973 war and the 1987 Palestinian intifada – became associated exclusively with the Arab 'other'.[26] It was so severely and negatively Arab in Israeli-Jewish eyes, that when right-wing extremists wanted to protest against Prime Minister Yizthak Rabin in 1995, following the Oslo Accords, they disseminated two photoshopped images of his

---

26   The kūfiyyā and shawls made from the kūfiyyā fabric are still used by religious-Zionist settlers, in particular women settlers, in the West Bank. We shall return to this point in the fourth chapter.

Figure 2.5   Ben-Gurion in Petra 1935
Source: Courtesy of the archive of Eilat.

face: as a SS Nazi officer and as an Arab 'terrorist' wearing a black-and-white kūfiyyā. This kūfiyyā, interestingly, which was first encountered and admired by early Zionists, and which became a symbol of localness and boldness by members of the Palmach and by Prime Minister Ben-Gurion, had become in the mid-1990s a symbol of an Arab enemy seen as on par with Nazi Germany.

In November 2014, this process reached its peak, when the kūfiyyā took on the role of the ultimate 'inciter'. Arab-Palestinian MK (member of the Knesset) Basel Ghattas, from the Arab party of Balad/Tajamuʻ (lit. the National Democratic Alliance) spoke at the Knesset wearing a kūfiyyā around his neck. MK Ghattas did this in solidarity with the Arab-Palestinian residents of East Jerusalem in light of growing national tensions in the city. His act of wearing a garment that used to be worn by the greatest leaders of Zionism, by Jewish fighters and peasants alike, by

Figure 2.6   David Ben-Gurion and his wife Paula on a Trip to Eilat 1935
Source: Courtesy of the archive of Eilat.

the first pioneers and the early Zionist immigrants, and of course by the Bedouin and Arab-Palestinian indigenous population throughout the years (including by Arab MKs while giving speeches at the Knesset in the 1950s and 1960s) – had become in Israel in the twenty-first century a direct sign of 'incitement'. MK Miri Regev, a leading member of the Likud party, who also served at the time as the chair of the Knesset's Internal Affairs Committee, shouted at Ghattas during his speech. She was not concerned with what he said, but rather with what he wore. She later wrote on her Facebook account:

> What does MK Ghatas think to himself, standing on the podium of the Knesset and wearing a kūfiyyā around his neck??? Until when will Arab MKs continue to spit in our face and incite against the State of Israel??? I asked the judicial advisor of the Knesset to examine whether the provocative act of MK Ghatas to wear a kūfiyyā in the Knesset is legal at all or perhaps a severe violation of the regulations of our parliament. I think that in his miserable performance, MK Ghattas only wanted to throw another piece of wood into the bonfire … In his act, MK Ghattas proven to be one of those Trojan horses that use the podium

of the Knesset to represent terrorist organisations and should therefore find his place outside the Knesset.[27]

Where Are the Arab-Palestinians in Israel's National Symbols?

Symbols were initially used by the Zionist movements as means of recruiting and bringing together diverse groups of people. As such, these symbols needed to address the common denominators of these groups. In the diaspora this meant utilising a range of images, narratives and values taken from Jewish religion and culture as well as from the socialist ideologies many of the early Zionist groups adhered to. However, the encounter with the land of Palestine and with local Arab-Palestinians, and the early difficulties Zionist settlers faced,[28] required the tailoring of the Zionist message. What was now required were devices to transform the settlers into a nation and market Palestine as a lucrative destination for potential European-Jewish recruits. This entailed forging a strong sense of settler belonging to the land, and in line with other settler-colonial endeavours, transforming the settlers into the natives. This transformation occurred through a variety of mechanisms, among them the notion of the reclamation of the biblical land of Erets Yisraeli. As such the local Arab-Palestinian population and their customs, and in particular the Bedouin and the *fallahs*, served as role models of the perceived and imagined Jewish biblical past. They were imagined as vehicles that helped preserve Jewish traditions, culture and symbols, and to which the Zionist settlers now 'returned'. Additionally, their connection to the land on the one hand, and their perceived freedom on the other, were used as inspiration for settlers trying to forge a new Jewish identity in opposition to the diaspora. Therefore, the adoption of local symbols was seen as a natural and neutral development. In other words, to romanticise and imitate local Arab-Palestinians was not seen in any way as antagonistic to the goals of the Zionist movement; this was especially true in the early pre-state period. As such, many local symbols, from local flora and fauna to clothing items and images, were adopted and later adapted and transformed into Zionist ones, connecting the present settlers with a perceived common cultural, religious and national past and creating a shared temporality. These symbols, however, once adopted, also took on a life of their own and were constantly reinterpreted, transformed and re-evaluated by Zionist and Jewish-Israeli society.

---

27   Interestingly, and particularly relevant in light of Regev's comments, there have been a number of recent attempts to bring back the kūfiyyā. In an interview with *Haaretz* newspaper the designer of the new Jewish 'blue and white' kūfiyyā claimed that 'the Palestinian [kūfiyyā] is black and white, the Jordanian is red and white. Why can't we have an Israeli one in blue and white? We are part of the region and as it is, we'll remain part of it' (quoted in Zerubavel, 2008).

28   Many Zionist settlers who immigrated to Palestine during the first and second 'aliyas subsequently left Palestine.

However, as the tension between the two communities increased, the desire to resemble and imitate the Arab-Palestinians decreased. Arab-Palestinians, therefore, while they were still romanticised to some extent, were also seen as competitors for the same land. As a consequence and in light of the Zionist movement's adoption of new national symbols that signified their resistance to the Arab threat (such as the lion of Tel Hai), national symbols were gradually de-Arabised. Later, and as soon as the symbols became Israeli, they also ceased to be Arab. This meant the loss and neglect of extensive local traditions, customs and history. As a result, and through the act of remembering its Jewish and Zionist past, Israel also forgot and erased its Arab one. In other words, through adopting the Arab symbols, the Zionist movement, and later the state of Israel, contributed to the marginalisation and later exclusion and expulsion of the local Arab-Palestinians and the transformation of the Zionist settlers into the natives.

The importance of highlighting the inherent Zionist need to create unique Jewish-Israeli symbols, and, at the same time, emphasising the effort made by Jewish-Israelis to differentiate and distance themselves from the Arab and Oriental world, is an integral part of our analysis. It is also, however, an integral part of the conflict. For example, in our attempts to secure the rights to use a number of images for this chapter, in particular those connected to the tsabar and the Zionist pioneer, we were accused of being 'anti-Israeli' and 'anti-Semitic'. Our willingness to allude to an Arab dimension of Israeli/Zionist culture and images was seen by some as treasonous. Below is the email reply we received, after we explained what our project was about, from a Jewish-Israeli person from whom we requested permission to use a number of images he held the rights for:

> I understand [what the project is about] and I have decided that I am not interested in being part of any project that serves anti-Zionist interests. The images of the tsabar and the [Zionist] pioneer have no equivalent anywhere in the world, and cannot be compared to some Arabness or Mizrahi identity [*mizrahiyut*] at all. The attempt [at comparison] smells to me like an anti-Semitic undermining [of the state of Israel], so I wish to you and to the anti-Israeli academy in the UK all the best in your endeavours.

Chapter 3

Digesting the Nation: Arab Ingredients in the Making of the 'Israeli Kitchen'

What is the connection between [Israeli minister of foreign affairs] Avigdor Lieberman and this place? What does he know about this place? I can look at the sky and by the shape of the clouds can tell you if it is going to rain soon or not. I can look at the sea and tell you by the height of the waves if it is a good day to catch fish or not. I can go out to the nature, four or five days after the rain, and tell you which herb is going to grow first and where. What does Lieberman know? Does he know any of that? Of course he does not.

MK Muhammad Barakeh, Haaretz, 20 November 2014

In the summer of 2014, we undertook a food research trip across Israel in search of the DNA of 'Israeli food'. Our trip and subsequent conversations with Israeli and Arab-Palestinian chefs, food writers and academics provided us with additional insights that extended and stretched our research from historical analyses to present-day events and trends. These insights made the relationship between Israeli food culture and national identity and the Arab-Palestinian kitchen and identity very clear indeed. It became obvious to us that the historical dialectical relationship between the local Arab-Palestinian culture and the Jewish-Zionist culture that emerged as part of their encounter in Palestine – which was a relationship based on the changing power relations between the initially hosting and immigrating societies – was still present.

These ideas first hit us when we entered a franchise of the Al-Babūr restaurant.[1] The restaurant, named after the Arab term for 'evaporation' used by Arab-Palestinians to depict the first steam ships that came to Palestine in the nineteenth century, is located at the main entrance to the Arab-Palestinian city of Umm al-Fahm, and also has a branch in the outskirts of the Jewish town of Yokneam. When we spoke with the restaurant's proprietor and head chef, Hussam Abbas, we learned a great deal about the texture of that steamy pot called 'Israeli food'. For Abbas, an Arab-Palestinian citizen of Israel, Israeli food culture contains nothing original, and is just an ongoing process of picking and choosing from other food cultures. For him, this Israeli cherry-picking process was concentrated on the Arab and Arab-Palestinian kitchen to which the Jewish immigrants were exposed to in Palestine/Israel.

---

1   Al-Babūr (sometimes also El-Babor) which serves mainly modern Arab-Palestinian food, has a mixed Jewish Palestinian clientele, and presents itself as a Mediterranean restaurant.

Upon leaving Abbas's restaurants, his words continued to echo. They then hit us in a different way when we stood in front of another restaurant just opposite the road from Al-Babūr's Yokneam branch, called Diana. Diana, an Arab-Palestinian Nazareth-based restaurant chain owned by Dukhul Al-Ṣafadī (an Arab family name linked to the Arab population of Safed that before 1948 was a majority Arab-Palestinian city and that today is all but emptied of its Arab-Palestinian population) was initially named after one of the first cinemas in Nazareth: Diana Cinema. It was another example of the way in which Palestine was transformed: from an Arab cinema; to a restaurant in Nazareth, with a mixed Jewish and Arab clientele; to a branch in the outskirts of the Jewish town of Yokneam, in which the majority of clientele is Jewish. And there, in front of the restaurant in Yokneam, we realised that neither Arab Safed nor the culture of the city of Nazareth were celebrated, but instead a modern and flashy sign signalled the new state of affairs in Israel/Palestine, between its people, cultures and food products. 'Diana – An Israeli Grill Restaurant – Kosher' said the sign, bringing together the Palestinian-related name Diana, with the invented concept of 'Israeli Grill' and with a Kosher certificate, as a metaphor for a final proof for 'Israelisation'. This indicated how Jewish-Israeli dominance and hegemony in the country has managed to force 'compromises' on the Palestinians to ensure the flow of Jewish-Israeli customers and money into the restaurant.

Figure 3.1   Diana – Kosher Israeli Grill

The tension between politicisation and de-politicisation grew more and more apparent to us as we continued our journey. We were therefore not surprised that after exchanging pleasantries and a few introductory sentences at 'Izbeh restaurant in Al-Rāmah village in northern Israel, the head chef Habib Daoud looked at us and said:

> so what are you trying to say? That you are writing about Israeli food? Tell me your politics and I tell you what you are looking for .... There is no such a thing as "Israeli food", you need to acknowledge that you are writing about politics, and food is just another angle of the Israeli-Palestinian conflict.

The relationship between food and politics in Israel/Palestine was evident almost all around us, and it became clear how central Arab food was and still is in the imagination of the Israeli food culture. For example, this was evident when urged by an Israeli shopkeeper to buy touristic presents for friends abroad. One of the souvenirs suggested was a series of place-mats with the 'Best of Israeli Food Recipes' on them. The 'best of Israeli food' collection included: hummus, falafel, Israeli chopped salad, pita, and kebab on a cinnamon stick – all of which are found at the heart of the regional Arab food culture, whether Palestinian or Shāmī.[2] It became clear that these examples, and many others, did not depict an Israeli reality but presented how it has been and still is shaped and imagined. It was not a surprise, therefore, to read in the Israeli daily *Haaretz* (Halutz, 2014) an homage to Israeli food culture celebrating the most successful Israeli restaurant in Vienna. There, the Jewish-Israeli chef Haya Molcho, who owns a successful Middle Eastern restaurant, described to the newspaper, with great pride and with no hesitation, 'how we have brought the Austrians Israeli food'. She then went on to mention the 'Israeli' food items in the restaurant that included hummus, labnah (an Arab strained yoghurt dish), laffa bread (laffa means 'roll' in Arabic) which Molcho calls on her menu 'Israeli bread', Libyan ḥarāimi fish stew, pita bread and of course falafel. Strikingly, all of the food examples presented by the Jewish-Israeli chef are Arab and Middle Eastern in their origin, and therefore in doing so she directly connected the 'Israeli food she brought to the Austrians' to mostly Arab as well as Arab-Palestinian food cultures.

The Vienna example only confirmed what we had already realised about the Israeli gastronomic DNA. We travelled throughout the country and realised that when our food 'antennae' were on, we kept receiving additional examples that were either unknown, or more evidently unnoticed, to us before starting this political, social and historical culinary journey in Israel/Palestine. We gradually began to ponder more seriously the tension that lies in the interaction between Jews and Arabs more generally, and that of their food cultures more specifically. We gathered that it has obviously not been a one-way process, and not always a

---

2   *Shāmī* (or *shaami*) food refers to the culture of *bilād al-shām*, the countries of the historical region of Sham, which includes Palestine, Syria and Lebanon.

conscious one, and that perceptions, narratives, imagination, desires and needs have taken part jointly in shaping the boundaries and tastes of the emerging 'Israeli food culture'. Some of these processes were connected to the Zionist desire to merge and localise into the country and region; some were a way of erasing or blurring the Palestinian connection to these food items; some – especially after 1948 – were related to the Palestinian reaction to Jewish dominance; and some were aimed at weakening the Israeli hegemony by means of resistance (Ashqar winery, for example, launched their latest wine 'Iqrit', named after the village whose inhabitants were expelled by the IDF in 1948 as part of Israeli 'security needs' and the Palestinian *Nakba*, in order to raise awareness of their struggle and the unjust caused to them). We realised that the tension that is found in food, and what feeds the nation – on a physical and political level – has a capacity to tell us a more complicated story than that of 'food' only. It is a story that has taken us to the Palestine of the nineteenth century, to the Zionist-Palestinian conflict, and to the cookbooks and restaurants of modern Israel. The tension that MK Barakeh spoke about grew livelier as our journey unfolded. It was a journey which was not only about the food that comes onto our plates, but also a struggle about authenticity and over origins. It is a struggle over the land of Israel/Palestine, which surrounds both Lieberman and Barakeh, and over which the two people keep on fighting, and which also keeps on feeding the Israeli-Palestinian conflict.

Food and National Identity

On first impression, food, including the everyday and mundane acts of cooking and eating, might not appear an important or even an illuminating prism through which to study the subject of national identity. However, eating and cooking are not only physical activities done in response to a bodily need, they are also means through which individuals and groups express their identity. It is probably a truism that what and how we eat conveys a lot about who we are to ourselves and to others (Mintz, 1985), in particular, our choice of cooking styles, ingredients and diets. Several writers, among them anthropologist Claude Lévi-Strauss (1986) and literary and social theorist Roland Barthes (in Counihan and Van Estrik 2008: 31), have gone so far as to suggest that particular food habits, manners, diets and tastes reflect the structure and culture of particular societies and even nations. Additionally, beyond the issue of culture and identity, food is also an important economic commodity that is directly tied and related to issues such as policy-making, national security and sovereignty. This dialectical relationship, between food as a necessity of everyday life and as an avenue through which individuals and groups express their identity, and its importance to the nation-state as a commodity, source of security and an important element in its policy-making, make it an exceptionally useful tool through which to examine the nation, the nation-state, and national identity and culture.[3]

Despite its apparent importance to issues of security and sovereignty, our main interest in this chapter is in food as a societal and cultural product. As such, the chapter focuses on the creation, maintenance and reproduction of an Israeli national food culture. By using the term *food culture* 'we do not simply mean a particular diet, but rather we include the manners and methods in which food is grown, produced, prepared, commodified, consumed, and perceived by a particular society' (Mendel and Ranta, 2014: 414). Our aim here is not to argue that Israel, or for that matter any nation, has a homogeneous food culture, but rather that it has particular and distinct food trends, narratives, items, images and methods of preparation that can be seen as part and parcel of the nation's identity and culture. We acknowledge, though, that this might not be apparent to Jewish-Israelis engaging with food in their everyday lives.

What is of particular interest to us is the way in which particular food items, cooking styles and diets relate to prevailing and accepted concepts, images and ideas of the nation. A number of writers have pointed out the importance of specific food items that have come to be associated with particular nations and which are used as a means of imagining, maintaining and practising the nation. Avieli (2005) and Rogers (2003), for example, provide evidence of the importance of rice cakes to the imagination and practice of a Vietnamese national identity and beef to British national identity respectively. In this regard, and as demonstrated in the previous chapter with regard to the prickly pear and the Jaffa orange, food items and practices can serve as symbols of and methods of engagement with the nation and national identities in line with Billig's idea of 'banal nationalism' (Palmer, 1998). The meaning attached to food items, however, does not remain static; it is constantly evolving and changing. According to Narayan (1995), food items and the meaning attached to them can be manipulated, transformed and even 'fabricated' in order to create new meaning and to reinterpret social relations. The ability to reinterpret and transform the meaning attached to food has also been used in many national settings as a way of unifying and bringing together different groups under the banner of the nation (Appadurai, 1988; Montanari, 2004). This relates directly to the ideas put forward by Edensor (2002) regarding the importance to the imagination of the nation of incorporating regional, religious and ethnic differences into the greater national identity. Food culture thus serves as a useful tool for bridging social difference on the one hand, but also for drawing and highlighting national boundaries on the other. Additionally, and more specific to our own case study involving Israel and Palestine, Ranta (2015) has shown that the transformation of meaning through the adoption and appropriation of

---

3   For more on the relationship between food and national identity, see: Ronald Ranta, and Atsuko Ichijo. *Food, National Identity and Nationalism: From Everyday to Global Politics* (Basingstoke: Palgrave Macmillan, 2015).

indigenous food items and traditions has been a widely occurring phenomenon in settler-colonial societies.

All of this demonstrates that studying national food culture can provide a useful means of accessing and highlighting 'social and political processes, watersheds and beliefs' and that food can serve 'as a historical recorder of much louder events, developments and decision-making' (Mendel and Ranta, 2014: 414). In this respect, and as argued by Bell and Valentine, the history of a nation's food culture is also the history of the nation itself, with food acting as a recorder of episodes of 'migration, trade, and exploration, cultural exchange and boundary-marking' (1997: 168–9).

The Importance of Food to Israeli Identity and Society

The relationship between food and Zionism is not at first apparent, after all Israel does not have a clear and identifiable national cuisine[4] and like many other settler-colonial and/or immigrant-based states, has a food culture which reflects a wide range of influences. However, and as demonstrated by several Jewish-Israeli writers, for example Avieli (2013), Hirsch (2011), Raviv (2002) and Rozin (2006), food culture and specific food practices and items have played an important part in the construction and representation of a Jewish-Israeli national identity. Additionally, the Zionist leadership during the Mandate period and the Israeli government since independence have used food as a way of promoting Jewish, and later Jewish-Israeli, food produce, while delineating and demarcating communal and national boundaries, in particular in relation to Arab-Palestinians. These policies were epitomised by the pre-state Zionist campaign of '*Totseret Haarets*' (lit. produce of the land), which emphasised the purchasing of Jewish-only produce at the expense of Arab-Palestinian and British produce, and the modern Israeli campaign of '*Konim Kaḥol Lavan*' (lit. we buy blue and white, the colour of the Israeli flag) to promote Israeli produce.

Writing for the Hadassah newsletter in 1940,[5] Sulamith Schwartz explains the logic behind the 'Totseret Haarets' campaign:

> The more we use the products of Jewish fields and factories, the more we encourage the development of Palestinian Jewish industry and agriculture, thus creating room and work for tens of thousands of new immigrants, strengthening the Palestinian Jewish economy, making it sounder and more self-reliant. (quoted in Raviv, 2002: 77)

---

4   A number of our interviewees even went so far as to state that there is no such thing as Israeli food, and that what is described as 'Israeli food' is in fact Jewish, Mediterranean, Palestinian, Arab or Middle Eastern food.

5   Hadassah is the Women's Zionist Organisation of America. During the Mandate period it was active in providing health and welfare support to the Jewish population.

The Zionist movement and its associated organisations, such as the Women's International Zionist Organisation (WIZO), used particular notions regarding food consumption, food practices and hygiene in a (not always successful) attempt at fashioning and transforming the immigrants based Jewish population into an 'ideal' and 'European' Zionist nation (see, for example, Hirsch, 2014; Rozin, 2006). This can be clearly seen in the introduction to the popular first Zionist cookbook, *How to Cook in Palestine* (Meyer, 1937), which was published by WIZO and which was continuously in print well into the 1950s:

> It is time that we, the women, try with greater vigour than before to free our kitchen from the Diaspora tradition that has clung to it ... and consciously replace the European cuisine with a healthy Israeli one .... This is one of the most important mechanisms for growing roots in our old-new homeland. (1937: 8)

The ideas presented by *How to Cook in Palestine* regarding how and what to eat were directly linked with the official 'Totseret Haarets' policy of promoting the purchasing of Jewish produce. This is explicitly stated by the author in the introduction. Meyer informs the reader that the great variety of recipes provided are based on the use of 'Totseret Haarets' agricultural and preserved produce. In addition, on page 26 an advertisement targeting Jewish housewives appears in English and Hebrew:

> To the Jewish housewife: Serve Totseret Ha'arets food only. In this way you will help strengthen the economic foundation of Erets-Israel. Take active part in promoting Totseret Ha'arets.

The importance of food to Jewish-Israeli national identity is also evident in the specific Jewish-Israeli national symbols which are food related. In the previous chapter we discussed the importance of the Jaffa orange and the prickly pear/ tsabar to the construction and maintenance of Jewish-Israeli identity, but there are other examples, such as the depiction of Israel as the land of milk and honey – a reference to the relationship between the land and the Bible. Perhaps one of the most well recognised Israeli food-related symbols is that of the pita with falafel. As shown in the Eldan rental car advertisement, falafel has become closely associated with the state of Israel and Jewish-Israeli food culture. So close is this association that one of Israel's most recognised postcards is of a pita bread with falafel balls and a flag of Israel, titled 'Falafel: Israel's national snack' (see Figure 3.1). 'The flag on the "mountain" of falafel looks almost as one put on a conquered destination, be it the US flag on the moon or another on the peak of the Everest' (Mendel and Ranta, 2014: 415). What is striking about this image is that a traditional Arab sandwich, popular across the whole of the Middle East, has become so closely identified with the Jewish-Israeli nation and with the state of Israel.

Figure 3.2   Falafel: Israel's National Snack

The strong relationship between Zionism, Israel and food is also evident in the importance attached to agriculture in economic and political life on the one hand, and the Zionist image and construction of the new Jew on the other. Pre-state Zionist leaders placed an emphasis on agricultural work as a means of physically and spiritually liberating Zionist settlers en route to creating a new nation. This included the establishment of collective agricultural settlements, for example the kibbutz, and the elevation of agricultural work and produce. This historical emphasis is still very visible in the symbols and emblems of many Israeli cities and towns that pay homage to the role of agriculture and contain images of rural life and agricultural tools and produce (see Figure 3.3 which shows the coat-of-arms of two Israeli cities).

An additional manifestation of the relationship between Jewish-Israeli national identity and food is the importance of religious-Jewish dietary laws and regulations (*kashrut*). Even though there is a great deal of debate regarding the importance of and adherence to Jewish religion in daily life, a majority of Jewish-Israelis follow kashrut laws to some extent. A recent survey by the Israel Democracy Institute found that over 60 per cent of Jewish-Israelis observe kashrut dietary laws and religious regulations regarding food consumption during Jewish holidays (Arian, 2014). This means that there are clear boundaries, in many areas related to food production, preparation and consumption, between Jewish-Israelis and non-Jews living in Israel. This parallels the importance kashrut laws had and still have to some extent in the diaspora, where they act 'as a barrier to free intermingling with non-Jews, fostering exclusiveness and separateness and ensured the perpetuation of an identity and a way of life' (Roden, 1999: 17).

Lastly, to fully appreciate the importance of food to Jewish-Israeli national identity and culture, one needs merely to spend a day or two travelling across the country. Food permeates most aspects of daily life in Israel. As we argue, as prevalent as political debates are in everyday life in Israel, food plays an important and similar role. From heated discussions over the price of food – in recent years the cost of specific food items, such as cottage cheese, have sparked protests that brought out hundreds of thousands of Jewish-Israelis to the streets – to passionate arguments over 'the best place to eat humus', food is a constant and dominant topic of conversation.

Figure 3.3   Coat-of-arms of the cities of Rishon LeZion (left) and Afula (right)

Israeli Cookbooks

For the purpose of researching for this book we read through and examined Israeli and Zionist cookbooks that purported to discuss or represent Israeli cooking and or food, from the 1930s – the period in which the first Zionist cookbooks were published – to 2014. These cookbooks, especially the early pre-state and post-state ones, provide a good indication of the dominant discourses used by the Jewish community in Palestine and later Israel. Many of the early cookbooks, while they were addressed mostly to housewives, tried to educate their readers with regard to gender, health and nutrition issues, but also with regard to expected national behaviour and values.

Reading through Zionist and Jewish-Israeli cookbooks it is evident that Israeli food culture has been influenced by a wide range of food traditions, processes, events and actors. Among these, several particular influences stand out. First, it is clear, especially with regard to cookbooks written between the 1930s and the 1950s that the Zionist settlers, especially the urban class, by and large did not try to create a new food culture but instead sought to replicate the food cultures they grew up on and knew from Europe. These, however, had to be adapted to the conditions and climate in Palestine and to the produce that was available. In other words, the early cookbooks were inspired by European and European-Jewish food cultures, but had to go through a process of adaptation to the climate, conditions, products and produce in historical Palestine, and later in Israel. As a consequence, many of these cookbooks stated the need to learn about and use the 'new' local ingredients (see, for example: Cornfeld, 1949; Meyer, 1937).

The second dominant influence on Israeli food culture comes from the growing science, today sometimes perceived as a 'trend', of nutrition. Several local Jewish and Israeli cookbook writers, for example, discuss the importance of balanced and healthy diets, in line with the balanced and healthy Jewish-Hebrew person ('the new Jew' or 'the non-diasporic Jew') they wanted to create in historical Palestine, later Israel.[6] The last important influencing factor discussed with regard to Israeli food culture is the contribution made by Jewish immigrants from Europe, North Africa and the Arab world.

Yet one of the most striking elements about Zionist and Israeli cookbooks is not the influences they discuss, which are varied and many, but the ones they fail to mention. With the exception of a small number of recent cookbooks, which will be discussed below, Zionist and Israeli cookbooks do not discuss and or provide any indication of an Arab and or an Arab-Palestinian element, contribution or influence on Israeli food culture.[7] This omission, by design or by default, of an Arab and or an Arab-Palestinian contribution to and influence on Israeli food culture is manifested through a number of different layers. The most obvious omission is the representation of Arab and Arab-Palestinian food as Mizrahi-Jewish food (see, for example, Cornfeld, 1949; Meyer, 1937; Peretz-Rubin, 1987; Sirkis, 1975). Through its association with Jewish-Mizrahi food, Arab and Arab-Palestinian food is connected to Jewish traditions and transformed into becoming

---

6   For a discussion of the importance of science to early Zionist and Israeli food writing, see: Ofra, Tene [in Hebrew] *Thus You Shall Cook! Readings in Israeli Cookbooks* [in Hebrew] (MA thesis, Tel Aviv University, 2002).

7   It is important at this stage that we clarify what we mean when we use the term *Arab-Palestinian food culture*, a term we neither intend to essentialise nor simplify. Our main argument here is that Zionist settlers first learned about and incorporated the food culture they encountered in Palestine through Arab-Palestinians. In line with Khalidi (2010), this does not mean that Arab-Palestinian food culture and identity are static, homogeneous and primordial. When put into context, we view food as a societal and cultural construct, and as such we 'take it for granted that Arab-Palestinian food culture was not created in a vacuum, but was constructed through a dialectical process influenced by the Middle East region, its people, visitors and rulers' (Mendel and Ranta, 2014: 418–19).

Israeli. The Mizrahim, therefore, served as a 'flashcard' that helped to prove the alleged non-appropriating nature of the Israeli appropriation of Arab and Arab-Palestinian food, and in its positioning of Arab food as part and parcel of the Israeli 'food culture'.

Another manifestation of this omission is the depiction of Arab and Arab-Palestinian food as Jewish biblical food (see, for example, Bar-David, 1964); some Israeli writers, for example, Shalev (2001), have even gone so far as to claim that hummus is in fact a Jewish biblical dish. This particular account represents Arab-Palestinians as vessels that allowed for the preservation of ancient Jewish biblical heritage. Rather than adopting and appropriating local Arab-Palestinian food items and traditions, Zionist settlers returned to or rediscovered 'their' own food history, in a similar fashion to their return to 'their' land. In addition, there are some writers who simply present Arab and Arab-Palestinian food items as belonging to a Jewish-Israeli food culture (see, for example, Zimenavoda, 1981).

This anomalous phenomenon somewhat surprisingly also includes cookbooks written by Arab-Palestinian citizens of Israel for the Jewish-Israeli market. In these cases, Arab and Arab-Palestinian food is presented either as a foreign food, in a similar fashion to other world cuisines, which have little bearing on or relation to Jewish-Israeli food culture, or as ethnic *Erets Yisraeli* food. The last category represents Arab food as part of the mixture of various ethnic groups in a perceived multicultural Israel. The terminology in these cases (Abbas and Rousso, 2006; Abu-Ghosh, 1996; Hinnawi, 2006) reflects the Jewish-Israeli hegemonic culture. The Arab-Palestinian food is described as 'Erets Yisraeli Arab food', 'Erets Yisraeli culture', and 'Galilean food' and the words Palestine and Palestinian are always omitted.

The historic omission of the Arab and Arab-Palestinian contribution and influence appears odd when one examines the importance of Arab and Arab-Palestinian restaurants, food terminology, dishes and culinary items in Israeli everyday life and culture. For Israelis, buying Arab food items such as pita bread, hummus, tahini, majadra and labnah at their local supermarkets, is seen as an integral part of Israeli culture, taste and daily life. Strikingly, even eating out in Arab-Palestinian restaurants (such as shawarma at Ḥazen's in Haifa, falafel at Mishel's in Wadi Nisnas, pastry products at the legendary Abulafia bakery in Jaffa established in 1879, hummus in Abu Gush village, and lamb's throat stuffed with minced meat and pine nuts in Al-Babūr restaurant in Umm al-Fahm) has been appropriated into and contextualised as part of the Israeli food culture (Mendel and Ranta, 2014: 420). The importance and contribution of Arab and Arab-Palestinian food is also evident in the way in which Jewish-Israeli chefs discuss Israeli food. When 60 leading Jewish-Israeli chefs were asked by ʿAkhbar *Ha-ʿIr* magazine (Golan, 2008) to state what Israel's national dish was, half of them mentioned food items that are mostly associated with an Arab and or Arab-Palestinian food culture. The food items identified by the Jewish-Israeli chefs included falafel, hummus, *musabaḥa* (hot chickpeas with tahini and lemon), aubergine with tahini (known in Arabic as *babba ghanoush*), and even 'Arab salad'. The association of Arab food items as Israeli food items was further corroborated by the Jewish-Israeli chefs we interviewed.

The omission of an Arab and an Arab-Palestinian element in Israeli cookbooks either implies that their contribution and influence was marginal, and that most of the dishes and traditions discussed above were introduced by Mizrahi Jews, or, as we argue, that for political, commercial and ideological reasons the Arab and Arab-Palestinian contribution was forgotten or censored. As we will demonstrate below, Israeli food culture evolved and was shaped by the Zionist encounter with the Arab-Palestinian people and their food culture. This encounter produced a dialectical relationship based on imitation, adaptation and de-Arabisation. In other words, the case of Israeli food resembles in many ways other Israeli national and cultural products.

Food in Palestine before Zionism

Zionist settlers often refer to the society they established in Palestine as the 'new yishuv'. This term was used by the settlers to distinguish themselves from what they termed the 'old yishuv': the pre-Zionist Jewish community in Palestine. In examining the contribution and influence of Arab-Palestinian food on Israeli food culture, it is important that we first look at what was eaten in Palestine prior to the arrival of the first Zionist settlers.[8]

The idea of creating the 'new Jew' was also connected to the early Zionist settlers' encounter with an indigenous population in Palestine. The 'new Jew' attitudes were framed, therefore, not only towards Arab-Palestinians but also towards the local Jewish population. The Jews of the 'new yishuv' perceived the pre-Zionist Jews in Palestine in an equivocal way: on the one hand, as another 'proof' of the continuous Jewish existence in Erets Yisraeli, but on the other hand as a degenerate, lazy and dependent community. The Jews of the 'old yishuv', however, viewed themselves as part of the Ottoman order in which they lived, and part of the Arab-dominated space they inhabited. They lived in mixed cities, spoke either fluent or basic Arabic (Kosover, 1966; Spolsky and Shohamy, 1999: 139) and adopted some of the local Arab customs.

With regard to *food culture*, and according to the Jewish food encyclopaedia (Marks, 2010), diffusion of food knowledge, traditions and practices was common between Jewish and neighbouring communities in the Middle East as well as in Europe. Jewish communities, as part of the countries they lived in, adopted, modified and reinterpreted local and existing bodies of culinary knowledge in line with Jewish dietary rules and customs. In addition, migrating Jewish groups brought with them and disseminated their culinary knowledge in their new environments.

---

8   For further reading, see Mendel and Ranta (2014). This section is largely based on our article.

This close resemblance between what was considered 'local' and 'Jewish' food, which was often the same, or at least influenced dialectically, was also present in Palestine among the 'old yishuv' prior to the Zionist emigration and colonisation. Ben-Naeh (2005) provides examples of customs common to middle and upper-class Jews and Muslims in the nineteenth century, which include, among other things, the drinking of a specific kind of black coffee, eating similar sweetmeats, and smoking *shisha* (water pipe). The influence of Arab-Palestinian food culture on the Jewish community is evident among the Sephardic and Ashkenazi communities alike, the latter, as mentioned earlier, even adopted to its Yiddish language in Palestine Arab food-related words, such as *kusalakh* (from Arabic *kusa*, meaning courgette), *keftalakh* (from Arabic *kufta*, meaning meatballs) and *zeytunes* (from Arabic *zaytun*, meaning olives) (Kosover, 1966: 245–58).

Discussing the Castel family (a Sephardic family of Spanish descent) in the early twentieth century in the Bukharan neighbourhood in Jerusalem, Sachar (1961: 23) describes the diffusion of local Arab customs into the Jewish home. He depicts how the family adopted:

> Arab, Turkish and Bukharan dress, Arab food and Arab furnishing. ... The family chairs were Arab wickerwork, and their tables were inlaid with Damascene carvings ... the coffee they drank was Turkish; their vegetables were Arab eggplant, or churned humus and techina [sic.] ... and their meat was grilled in oil, Arab fashion.

The following account of the Shlush family's food traditions, towards the end of the nineteenth century, illustrates the central place of Arab-Palestinian food in the pre-Zionist Jewish community. The Sephardic family, who arrived in Palestine from Morocco in the mid-nineteenth century, lived in the city of Jaffa. According to the family memoirs, their house contained a large central courtyard where the main oven (*ṭābūn*) for baking pita bread stood. Dinner, which was the main meal of the day, was served in a large clay dish and diners, sitting on the floor, would serve themselves using pita breads. Small dipping plates with olive oil, olives and labnah would also be provided. After dinner, sweet delicacies, such as dried dates and figs, *baklawas* (according to the same memoirs, the Damascene sweets were especially prized), *rahat lokum* (Turkish delight) and *tamar-hindi* (tamarind), would be served alongside strong Arabic coffee. Breakfast and lunch would normally be lighter and consist mainly of pita breads with *za'atar* (a herb with similar taste to thyme), olive oil, labnah, hard-boiled eggs, and olives. The family diet included little meat or fish, but on special occasions they would also serve stuffed vegetables, in particular courgette (*kūsā maḥshī*) and vine leaves (*waraq al-'inab*). In addition, the family also preserved the tradition of eating couscous, which Muslim and Jewish immigrants brought with them from North Africa, as well as eating *ḥamin* (a traditional Jewish stew) on Saturdays (Shlush, 1991).

The stories of the Castel and Shlush families are but two examples of the Jewish food culture reality in Palestine at this crucial moment in history, on the eve and in the first few decades of the Zionist immigration to the country. In many ways, the story of Jewish families who lived in Palestine during that period, who adopted or modified local-indigenous food traditions and items either according to necessity or taste, is similar to other historical case studies involving settler, colonial and/or immigrant communities.

Zionist Arrival and Zionist Return

The overall aim of the Zionist movement was not only to establish a Jewish state in Palestine, but also to create a new Jewish society and identity. The desired new 'Jewish' identity was to be constructed in opposition to the diaspora Jew, and to some extent also the Jew of the old yishuv. In terms of food, this meant rejecting the meat- and fish-heavy and slow cooking traditions of Eastern Europe for a perceived healthier diet based on fresh fruit, vegetables and dairy products (Almog, 2000). According to Claudia Roden 'the early pioneers and the first immigrants from Europe ... were happy to abandon the "Yiddish" foods of Russia and Poland as a revolt against a past identity and an old life ... and foods that represented exile and martyrdom' (1999: 175).

This was directly linked to the Zionist focus on agriculture and manual labour as a way of reviving the Jewish spirit and constructing a new Jewish identity.[9] As Almog (2000) demonstrates, this 'new Jew' in Palestine was portrayed as the antithesis of the 'Jewboy of the Diaspora'; the former was raised in a 'rural environment', had an 'active lifestyle' and was 'physically uninhibited'; he was strong in body and in spirit. This rejection of their old identity necessitated a search for new models upon which to base their new identity. As such, and despite their partial dismissiveness of the urban Arab-Palestinian culture and traditions, the images of the Bedouin warrior and Arab-Palestinian *fallah* served as worthy models to emulate. They were imagined and romanticised as modern biblical forefathers and seen as a source of inspiration and of imitation (Yaacov, 1981).

It was not just the images of locality that the new settlers required. Like other settler-colonial endeavours, early Zionist settler attempts to survive and flourish in their new environment were to a large extent dependent on two factors. First, the support and funding they received from Zionist groups, organisations and several Jewish-Zionist philanthropists. The financial and logistic support provided by these, in particular by Baron Rothschild, helped the early settlements avoid calamity (Sacher, 1961). Second, and just as important, the settlers survived and flourished due to the support, trade, and knowledge they gained from the local Arab-Palestinians. This reliance on local Arab-Palestinians was particularly extensive for first *'aliya* members, but continued well up to the 1948 war. The use of, to varying degrees, local agricultural techniques, bureaucratic knowledge

---

9   Parts of the section that is to follow are based on our earlier articles Mendel and Ranta (2014) and Ranta (2015).

(especially with regard to the Ottoman structures of power) trade and labour meant that the new Zionist settlers were dependent and reliant on local population (Yaacov, 1981).

As a result of their reliance on local Arab-Palestinians and constant contact between the communities, settlers began to adopt local customs, some even going so far as to learn Arabic. These 'new Jews' did not take to some of the local food items and traditions; for example, Zionist food writers were still imploring Jews to try aubergines, olives and local herbs and spices well into the 1940s and 1950s (see, for example, Meyer, 1937). Nonetheless, it made little sense to try to reinvent the culinary wheel in Palestine. Many of the food items desired by the settlers were hard to acquire and or not available. For example, in the first decade of the twentieth century, David Ben-Gurion wrote that 'there is no milk in the summer, it is almost impossible to find butter, bread is not the choicest, meat is a bit more expensive'; he went on to say that although there were local products which were cheap and easy to get, one first had to learn how to use them (Shafir, 1996: 70). In this respect, the settlers needed to adapt to the new surroundings by imitating and adapting the local food culture. The local Arab-Palestinian population knew what grew well and how best to use local produce.

The imitation and adaptation of local food was not only done out of need. Zionist settlers, for example, expressed their joy and fascination at learning to bake their 'own' pita bread and eating it in the fields with olive oil and labnah. They described the (somewhat) exotic experience of eating in Bedouin tents as akin to rediscovering the Bible (Raviv, 2002). The adoption of these food items and practices was further enabled by the Zionists' emphasis on socialism, connection to the land, the rejection of European lifestyle and culture, and, as mentioned before, a general attitude of negation of the diaspora. According to Roden (in Kantor, 2002), many groups within the Zionist movement, therefore, showed little interest in the 'jewels' of Palestinian cooking, but were attracted to the 'humble street food' that probably represented for them the roughness of their new habitat and their connection to it. Raviv illustrates this anti-bourgeois trend in the adoption of falafel as 'a quick, no frills, affordable, and satisfying food' (2003: 20). Gvion and Hirsch argue that the main motivating factors behind the adoption of certain food items and practices were their functionality, simplicity and how easily they fitted in with Jewish dietary laws (most of the food items adopted were *pareve*, lit. 'neutral' in Yiddish, which means, in the Jewish culinary context, they contain neither meat nor dairy products).

We argue that the fascination with, and the romanticisation and imitation of the local Arab-Palestinian, together with a desire (overt or covert) to become a political *replacement* to this people, were evident processes throughout the Zionist immigration waves to Palestine, and indeed to this very day. However, it is also clear that the emphasis accorded to these different processes changed with time. As the conflict between the two national groups grew, and the Jewish numbers and desire to establish a separate Jewish society and economy increased, the levels of romanticisation and fascination decreased and the aim of replacement increased in strength. In other words, as the conflict heated up, the focus changed from imitation and adaptation to appropriation and denial, a process we characterise as de-Arabisation. This process and dialectical relationship is evident in what is considered today as Israeli food. The added value that was attached to Palestinian food and food culture was its Israelisation and de-politicisation, two processes that exist on different levels and that helped distance the food and its culture from Palestine and Palestinians, thus making it a 'natural' part of the Israeli experience.

Apparently, and as the not-so-surprising Israeli adage says, 'life is in a pita bread' (lit. *ha-ḥayim be-pita*). 'Israeli food', or what many Israelis considered as 'Israeli', frequently comes in pita bread, the round bread which Zionist immigrants probably tasted for the first time when they encountered the Palestinian kitchen. Interestingly, despite the fact that Western countries imported the *pita* as an Arab-oriented bread and cultural product, the name is taken not from Arabic but from Aramaic. The name *pita* is closer linguistically to the Hebrew *pat*, meaning food items made of dough. Interestingly, it is a unique example of an Arab-oriented cultural (food) product that is mostly known by its Hebrew-oriented name, which – paradoxically – is difficult for Arab speakers to pronounce (this is because of the lack of the consonant 'p' in Arabic). In many ways the story of the pita is similar to that of the Jaffa orange case study that was discussed in the previous chapter. It demonstrates the impact of Israeli appropriation, but, even more importantly, the far-reaching effects of mass production, export and trademarking, which have all contributed to the Israelisation of an Arab food item, within Israeli society and also on a global level.

There is therefore an almost ironic meaning to the fact that most of what Israelis consider as 'Israeli food' comes inside an Arab-Palestinian cultural artefact. Paradoxically, the act of Israelising food, as we argue in this book, comes through its Palestinianisation – many of the foods that need to be nationalised as 'Israeli' have to come into and through the pita. From the famous Israeli *schnitzel be-pita* (chicken escalope in a pita), through *shakshuka be-pita* (egg poached in spicy tomato sauce in a pita) everything that needs to be considered as '*kosher*' from the Israeli national point of view goes into the pita. Perhaps the most extreme version of this phenomenon can be found in one of Tel Aviv's most successful delis ('*Ha-Miznon*' – the 'Kiosk'), where an Israeli TV celebrity chef, Eyal Shani, sells fancy food in pita. There, you can have for main course 'pink shrimps in a pita' and for dessert 'bananas and chocolate spread in a pita'.

One can argue that there is a lot of the 'new Jew' spirit in Shani's place, as well as in the pita-oriented street food culture of Israel. The eating of food with no cutlery, the baldness of eating in the street, and in a local, Middle Eastern bread, all involve the desired negation of the diaspora and its lifestyle. This 'pita-culture', in one way or another, also corresponds with the elevated character of *chutzpah* which seems to represent Israeli 'rough', 'impolite', 'straightforward' spirit (Gitelman, 1977: 82). One can argue that the simple act of eating in the street, standing up and perhaps more graphically, with tahini dripping from one's mouth – is almost a replica of the image of 'Srulik'. All in all, and perhaps in an unconscious revolt against the home- and slow-cooked food of Eastern Europe, Israeli culture identified as 'most local' those food items that are eaten outside in the street, that have a Middle Eastern 'envelope' and an 'Israeli' content. This new creation, we argue, captures the Israeli 'formula' that makes food 'local' and 'Israeli' and that brings about the desired 'new Jew' through an unconscious de-Palestinianisation of these very food items.

An example of this de-Palestinianisation can be also seen through one of the meals Israelis are most proud of: 'the Israeli breakfast.' On many occasions, one can experience Israeli pride in the local breakfast and the food it contains. The following is a description of the 'Israeli breakfast' as advertised on an Israeli leading tourist website (Go-Jerusalem):

> There is a reason why the hotels in Jerusalem go all the way with their Israeli breakfast in the menu. The Israelis know how to make the best continental breakfast in the world. … If you visited any place in Europe, you already know that their breakfasts might contain only a poached egg and some chopped vegetables. But with us you will get a buffet of respect: from salads, spreads, green omelette (lit. *ḥavitat yerek*: a fresh herb omelette), other kinds of eggs, yoghurts, labnah, smoked fish, seasonal fruits, hard and soft cheeses, croissants, breads, bagels, coffee, freshly squeezed juice and even shakshuka if you like hot dishes in the morning. Yes, there is a reason why it is called an Israeli breakfast.

The Israeli spirit that blows through the text is almost obvious. The Israeli breakfast is 'the best in the world'; it has lots of 'respect'; the Europeans' have only 'a poached egg and some chopped vegetables'; and only in Israel one can find this combination of 'modernity' in the Middle East, fresh and simple food with a sophisticated twist of variety. The 'Israeli breakfast' as a gastronomic representation of Zionism's founding father Theodor Herzl's idea of Israel as 'a wall against Oriental barbarianism', or of Israel's former Prime Minister Ehud Barak's description of Israel as a 'villa in the jungle'.

Deconstructing the notion of 'Israeli breakfast' further, though, uncovers the process of the Israelising of a Palestinian concept. First, the 'Israeliness' of Israeli breakfast comes from its pre-state Zionist origins – the Kibbutzim. There, the combination of socialism, agriculture and patriotism was the epitome of the 'new Jew', and it is no coincidence that the symbol of Israel's 'national cheese', the cottage cheese, resembles a kibbutz house. The kibbutz connection to the concept of 'Israeli breakfast' was highlighted by many, among them Torstrick (2004: 110):

> The typical Israeli breakfast owes its origins to kibbutz life. Kibbutz members needed a hearty meal to carry them through the day's work. A kibbutz breakfast buffet might consists of sliced fresh fruit (melons, oranges, apples, pears), sliced vegetables (cucumbers, tomatoes, peppers), olives, a variety of breads, yoghurts,

numerous cheeses or cheese spreads such as labnah, other kinds of eggs (hard-boiled, soft-boiled, scrambled, or poached), juice, and coffee.

One might wonder by whom were the Kibbutzim inspired when coming up with their 'hearty meal' that later become one of the symbols of the Israeli identity. The answer, as in many other cases, is found in the local Arab-Palestinian community, who, unsurprisingly, have a very similar breakfast. Among the similarities are the lack of meat, the emphasis on fresh fruits and vegetables, hard-boiled eggs and herb omelettes (known in Arabic as 'ijja), as well as olives, pita bread and labnah.

A third example of the process of de-Palestinianisation can be seen in the concept of the 'Israeli salad'. There might not be a more popular dish in Israel than the Israeli salad, sometimes also referred to as *Salat Katsuts* (lit. 'chopped salad'). The salad is based on chopped vegetables (normally tomatoes, cucumbers and onions), fresh herbs (mostly parsley but sometimes also mint) and dressed with olive oil and lemon juice. A recent book titled *Fresh Flavours from Israel* by Jewish-Israeli food writer Janna Gur (2008) states that 'Israelis must have their salad at least once a day'. The blog 'How to be an Israeli' suggests you must learn how to make the salad if you want to become an Israeli. It is an accompaniment to every meal, whether eaten at home or outside. In fact, you would be hard pressed to have a meal anywhere in Israel without it. No Israeli cookbook from the 1960s onwards is truly complete without providing a recipe for it. This is true with regard to those written for Jewish-Israeli and foreign audiences. To the unsuspecting viewer, the Israeli salad is the epitome of Israeli food culture: it is fresh, simple, healthy and symbolises the strong relationship the nation has with its agricultural produce. The salad, therefore, serves as both an internal and an external banal symbol of Jewish-Israeli identity.

What are the origins of the Israeli salad? Reading through literary accounts of growing up in Israel, the salad became a staple food product in the Kibbutzim mess halls, from there it moved to the Israeli army's kitchen and to Israeli homes. The fact that the salad is mentioned, mostly as a chopped vegetable salad by Israeli authors describing living in Israel in the 1930s and 1940s (see, for example, Amos Oz, *A Tale of Love and Darkness* and Meir Shalev, *Russian Novel*) demonstrates that the salad did not arrive with the waves of Jewish immigrants from North Africa and the Arab world after the formation of the state in 1948. It was also not prevalent in the diets of Central and Eastern European Jewish communities, from where most of the early Zionist immigrants came. On the other hand, there are a number of accounts, mostly by travel writers, such as Masterman and Grant in the early 1900s and that of the Mary Eliza Rogers as far back as 1865, which describe the preparation and consumption of a chopped vegetable salad in Palestine.[10]

---

10   See, for example, Elihu Grant, *The Peasantry of Palestine* (New York: Pilgrim Press, 1907); E.W.G. Masterman (1901) 'Food and its Preparation in Modern Palestine' *The Biblical World* 17(6): 407–19; and Eliza Mary Rogers, *Domestic Life in Palestine* (Cincinnati: Poe & Hitchcock. 1865).

Rogers (1865) describes the salad as accompanying most meals served by the upper and governing classes in Palestine, at the time part of the Ottoman Empire. In other words, either that early Zionist immigrants to Palestine independently invented the salad, or, as we argue hereafter, they imitated, adopted and later appropriated and nationalised an existing local custom.

Until recently the salad was always referred to as an Israeli salad. In recent years, however, several Israeli writers have openly acknowledged the salad's 'Arabness'. When asked what Israel's national dish is, chef Nir Tzuk claimed it was 'Arab salad', which according to him 'is craved by every person in the state'. In their recent cookbook *Lamb, Mint and Pine Nuts: The Flavours of the Israeli-Arab Cuisine*, Arab-Israeli chef Husam Abbas and Jewish-Israeli food writer Nira Rousso (2006) refer to the salad as an Arab salad. In an interview with the BBC, Jewish-Israeli chef Gil Hovav (2008) 'admitted' that 'this salad that we call an Israeli salad is actually an Arab salad, a Palestinian salad'. Additionally, many restaurants, especially Jewish-Mizrahi ones, have in recent years started to use term 'Arab salad' to describe in their menus the finely diced tomato-cucumber-onion salad. Others, like the Israeli restaurateur Sharon Shaked (2003) stated in jest that 'in truth, there is no such thing as Israeli salad – there is an Arab vegetable salad, but in the name of our independence why should we really care?'

Re-Arabisation of Israeli Food?

The Israeli late willingness to 'admit' the origins of the Israeli salad is a very interesting and illuminating case study. Until very recently there was hardly any recognition of the Arab and or Arab-Palestinian influence on Jewish-Israeli food. In recent years, however, there seems to be more opening and recognition, at times even celebration, albeit in a very limited scope, of Arab and Arab-Palestinian food in Israel. One of the clearest examples of this phenomenon is the rise of *hummusiyot* (singular: *hummusiya*) in Israel: a restaurant that specialises and serves mainly 'gourmet', made-from-scratch hummus.

Several of the Jewish-Israeli chefs we interviewed raised an important point with regard to the consumption of hummus in Israel. Though they accepted the Arab origins of the dish, they also claimed that the manner in which hummus is consumed and served in Israel is at times different and distinct from the Arab and Arab-Palestinian food cultures. In other words, they argue that hummus has been to some extent Israelised. One of the chefs for example asked: where else in the world do people eat hummus with warm mushrooms? The Israelisation of appropriated food is also found in other fronts; for example, it is common to be served schnitzel (breaded chicken escalope) with hummus in a baguette.

What is of interest to us is that the 'gourmet-isation' of hummus in Israel has gone hand in hand with the recognition of Arab and Arab-Palestinian hummus as being superior in quality and as being more authentic.[11] This point is made clear in a ground-breaking book on hummus written by Yehuda Litany and Naim Araidi (2000); they argue that the best hummus places in Israel are almost exclusively Arab-Palestinian. This recognition is also apparent in Israel's industrial and mass produced hummus industry. For example, one of Israel's leading salad producers (Tsabar) has been selling 'authentic' Arab hummus for several years. This includes its 'Kings of Hummus' series (see Figure 3.4), based on the recipes of prominent Israeli Arab-Palestinian chefs, which includes the '*musabaḥa*' (warm boiled chickpeas with tahini and lemon) of Big Samir; Big Samir's hummus with extra tahini; the superior hummus texture of Abu Marwan; and the spicy hummus of Abu Marwan. In search of the best hummus in the Middle East, the company also produced a special 'Nihad the Jordanian' line of hummus. There are a number of additional examples of this trend which include, among others, tahini, coffee, baklava and halva. Why are Israeli food companies promoting or 'inventing' authentic Arab chefs, products and recipes?[12]

Many of the historical accounts of romanticisation and imitation leading to the appropriation of the local Arab-Palestinian culture focus primarily on the images of the Bedouin and the fallaḥ, and mostly end with the 1948 war. However, this narrative only conveys part of the story. Jewish-Israeli fascination with the Arab and Arab-Palestinian culture in general, and food culture in particular, was never limited to the Bedouin and the fallaḥ, and did not end in 1948, in fact it has continued unabated, albeit in different forms.[13] This fascination has not been limited to specific food items such as humus, falafel and salad, but encapsulates a wide range of products, traditions and methods. We would argue that in recent years this phenomenon has manifested itself through three separate processes, which at times overlap. First, Arab-Palestinians and their food culture have continued to 'provide' inspiration and sources of emulation to Jewish-Israeli chefs looking to engage with ideas of

---

11   For a thorough account of the history of hummus in Israel and its recent 'gourmet-isation', see Dafna Hirsch (2011) 'Hummus is Best When it is Fresh and Made by Arabs: The Gourmatization of Hummus in Israel and the Return of the Repressed Arab' *American Ethnologist* 38 (4): 617–30; and Dafna Hirsch and Ofra Tene (2013) 'Hummus: The Making of an Israeli Culinary Cult' *Journal of Consumer Culture* 13 (1): 25–45.

12   One could add to this the recent expansion of several Arab-Palestinian restaurants and restaurant franchises into Jewish-Israeli towns and cities (for example, Al-Babūr and Diana); the celebration, in some food circles, of new Arab-Palestinian restaurants (for example, 'Izbeh, Sharābīk and Turquoise); the appearance of Arab-Palestinian chefs on Israeli cooking programmes; and the fact that last year's winner of Israeli's version of Masterchef (a cooking reality TV show) was Nuf 'Atamana-Isma'īl – an Arab-Palestinian citizen of Israel (the previous year's runner up was also an Arab-Palestinian citizen of Israel).

13   See, for example, Litani and Araidi's discussion of the importance of 1967 and Israel's occupation of the West Bank and the Gaza Strip to Jewish-Israeli and Arab-Palestinian food relations: Yehuda Litani and Naim Araidi [in Hebrew], *Not by Hummus Alone: Hummus, Olive Oil, References* (Tel Aviv: Dinur and Modan, 2000).

Figure 3.4   Tsabar's 'Kings of Hummus' Range: Big Samir's Hummus with Extra Tahini and the Superior Hummus Texture of Abu Marwan

locality and seasonality. Drawing on recent global food trends of utilising local, organic and wild ingredients, and looking to connect cooking to landscapes, nature and culture, Jewish-Israeli chefs and restaurants have recently started to cook with local herbs, cereals and plants. Instead of reinventing the wheel, the inspiration for these food items has come from their use in the Arab-Palestinian kitchen.

For example, Jewish-Israeli chefs and food writers have exalted the health benefits and value of the *freekeh* (smoked green wheat), describing it as the Israeli answer to quinoa and the hip grain of 2012 (Guttman, 2012). The fact that the cereal is and has been used by Arab-Palestinians and that the best place to buy it is in Arab-Palestinian markets is widely acknowledged. However, in a new-old twist on the de-Arabisation phenomenon, many of the Jewish-Israeli writers who provide recipes for *freekeh* also find it important to try and link the cereal to certain biblical passages (see, for example, Ansky, 2009; Gerty, 2007; Guttman, 2012).

Other obvious examples are to do with the Israeli fascination with Palestinian herbs. Za'atar is most famously *the* symbol of the Arab-Palestinian food culture, and also – following the Palestinian *Nakba* – a symbol of both the yearning for

Palestine and the hopeful return to it.[14] Za'atar is a mixture of herbs mostly based on wild oregano (*origanum syriacum*) that has a distinctive taste and smell, which is both wild and local, and which is believed to symbolise Palestine, in a modest, cultural, and simple way. As such, za'atar is mentioned in Arab-Palestinian short stories and novels as well as poems, and it is the one famous food item that Arab-Palestinians outside the homeland ask friends and relatives to bring with them. Salman Natour, a famous Arab-Palestinian writer, for example, in his book 'Safar 'ala Safar' (lit. 'A Journey over Journey') includes an unforgettable scene in which an Arab-Palestinian from Israel who travels to London is being body searched and sniffed by police dogs, as they all fear his za'atar is a forbidden drug being smuggled into the country.[15] Yet even though za'atar has such a strong connection with Arab-Palestinian national and cultural identity – and perhaps because it is has such a connection – one can find a mixture of feelings in Israel towards its 'origin'. And so, alongside some Jewish-Israeli food sources that explain za'atar's Arab context, others begin by mentioning its Israeli spirit, paraphrasing the za'atar herb to be 'the green Israeli'. Unsurprisingly, some even manage to find a biblical passage that 'relates' to it. Today the herb is widely used in the Jewish-Israeli food industry, can be bought in almost any food store in Israel and is often found as an accompaniment to many pastry products and salads.

Perhaps this straightforward inclusion of za'atar made the entrance of other herbs into the Israeli kitchen easier. We are referring to the fascination that began in Israel over the last decade or so – and as part of the international healthy trend towards healthy food – which has resulted in herbs such as 'akūb (*gundelia*), 'īlet (*cichorium*) and *khubāzah* (*malva*) gaining prominence on the plates of trendy Israeli restaurants and on the shelves of organic food stores. Perhaps these herbs' Arabic names, within a specific context – the rather 'lefty' organic food spirit – made them a more welcome addition in Israeli health stores. They were something very 'wild' and 'rough' and perhaps corresponded well with the international health trend towards 'raw food', 'going back to nature', or even a hidden take on the Paleolithic diet – again regarding Arab-Palestinians as noble savages, who can be romanticised and admired.

The second process has been the clear drive among food manufacturers as well as consumers towards producing and marketing more 'authentic' food products. A more 'authentic' product, which is seen to be based on artisanal traditions, is perceived to be of better quality. This drive is not only happening in the context of Israeli-Palestinian relations, but is part of a wider trend that has been amplified by the rise of globalisation. This need to add certain qualities, histories and traditions into particular products to increase their marketability can be seen in many different

---

14   Many regions in the Middle East have their own distinct type of za'atar, though it is closely associated with Palestine; Arab-Palestinian za'atar is based on particular types of wild oregano and thyme.

15   See, Salman Natour [in Arabic], 'Safar 'ala Safar', in Salman Natour (trilogy in Arabic), *Sixty Years: Walking in the Desert* (Ramalla: Dār al-Shurūq, 2009), pp. 216–19.

settings worldwide (Atsuko and Ranta, 2015). For example, in Israel this can be seen in the growth of Jewish-Mizrahi 'authentic' restaurants, cookbooks and food products; Jewish-Mizrahi communities reconnecting with what is seen as their historic culinary cultures and traditions. As part of this search for better and more 'authentic' food, Arab and Arab-Palestinian food, influence, traditions, restaurants and chefs have been acknowledged and at times even celebrated. However, while we argue that this development is welcome and opens up the possibility for better communal relations and understanding – a point to which we will return – it also needs to be contextualised.

Several of our interviewees argue that the recent acknowledgment of Arab and Arab-Palestinian food and influence is not part of a new process of inclusion and of creating new spaces for coexistence and cooperation, but rather a new and useful mechanism for subduing and domesticating a threatening space. In other words, and using Lisa Heldke's (2003) terminology, through Jewish-Israeli food culture hegemony the Arab-Palestinian Other is essentialised and imagined only as a source of several tasty food products. While Arab-Palestinian chefs are 'allowed' to show their skills on TV and publish their recipes through Jewish-Israeli publishers and websites, the space provided for them is limited to certain dishes and there is an implicit understanding this can only happen within the Jewish-Israeli hegemonic discourse. This inclusion does not indicate any recognition of the history, identity and political aspirations of Arab-Palestinians as a national group let alone any mention of the ongoing occupation. This means that the only way for such chefs and indeed food products to enter the Jewish-Israeli cultural space is under the banner of Israeli, Arab-Israeli and Galilean food.

Third, despite the recent recognition of Arab-Palestinian chefs, influences and food products, there is also an ongoing process of appropriation and de-Arabisation. This modern appropriation is done in the open and in a transparent manner. It reflects a Jewish-Israeli society that is more secure and confident in its political and cultural domination over Arab-Palestinians. The food products that are seen as useful and tasty are now openly imitated, adapted and appropriated. This is in line with Wilk's (2006) argument that food serves as a useful barometer for power relations, in that dominant societies are able to adopt and transform the food of others into their own. Let us give an example of this trend.

One of the best known and loved Arab-Palestinian desserts is *Knāfeh* (a cheese pastry with *kadaif* noodles soaked in sugar syrup). It is widely accepted and associated with the Arab-Palestinian kitchen in general and with the city of Nablus in particular. Over the years the dessert has always been acknowledged as Arab and, as such, featured in the dessert menus of many Israeli restaurants and coffee shops. However, in recent years the dessert has been more often discussed and presented as Jewish-Israeli rather than Arab-Palestinian.[16] This process has

---

16 See, for example: Ali Abunimah, 'Did You Know? Palestine's Knafeh is Now "Israeli" Too?' 3 June 2014 http://electronicintifada.net/blogs/ali-abunimah/did-you-know-

now moved one step further with the opening of a Jewish-Israeli restaurant in Tel Aviv dedicated to Knāfeh, where one could find, for example, savoury Knāfeh with shakshuka and Knāfeh with Belgian chocolate. The restaurant has been at the centre of a social media campaign and protest over the appropriation of the Knāfeh by Jewish-Israelis.[17]

We found another, though slightly different, manifestation of this phenomenon during Israel's recent Independence Day (Zaka'im, 2015), when the Israeli daily *Haaretz* chose musabaha (mentioned above), the Shāmi/Arab-Palestinian dish, as one of 'Israel's finest food for the Israeli Day of Independence'. In the article, a Jewish-Israeli chef explained that this is a 'Mediterranean dish' fit for the 'Mediterranean weather and ingredients', without ever mentioning that it has Arab and Arab-Palestinian connections, and that that name of the dish in Arabic means either 'to be praised' (due to its taste) or 'to be soaked' (as the chickpeas 'swim' in tahini paste). The article treats musabaha as just a word, with no origins and cultural connections.

Food for Thought: What Does This All Mean?

As we have shown throughout this chapter, food, because of its ubiquitous, mundane and everyday nature, is a useful tool for studying issues concerning national identity as well as long-term and large-scale social and historical processes. Though on the surface it sometimes does not appear to be an integral part of Israeli national identity and culture, as we have demonstrated, food culture has played an important role in shaping, maintaining and symbolising Israeliness and the Israeli way of life. As we discussed in the previous chapters, the Zionist construction of the 'new' Jewish identity necessitated the rejection and negation of their previous diaspora identity in parallel to the process of 'localisation'. This process was achieved, in a significant way, through the incorporation and emulation of the local Arab-Palestinian identity and culture. In terms of food, this meant changing the settlers' Eastern European diet to one more closely related to the local Arab and Arab-Palestinian food culture. This meant, for example, substituting a meat- and boiled vegetable-rich diet to one more dependent on salads and dairy products. The change was part of the settlers' desire to transform themselves into the native population and to localise their identity and culture. One could argue that it was

---

palestines-knafeh-now-israeli-too (accessed 10 June 2014), and Shiri Katz [in Hebrew], 'The Knafeh is (Not Always) Sweet' *Timeout Tel Aviv*, 24 November 2013.

17   When asked about the claims of appropriating an Arab-Palestinian dish, the owner of the Israeli Knāfeh restaurant stated that he could not understand why anyone would be angry with him. He explained that in his view food should be about bringing people together, not separating them, and that he should be judged only on the quality of the Knāfeh he serves. He went on to say that 'the nice thing about food, throughout history, is the way in which it migrates from place to place and changes over time' (Vered, 2014).

food, as a product without which the body cannot grow or function, which was so crucial for the creation of the Jewish-Israeli national body.

The transformation initially started with the fascination, adoption and imitation of local food cultural elements. However, as the conflict between Zionists and Arab-Palestinians intensified in the run-up to 1948, and the desire to create a separate Jewish economy and society increased, the initial fascination turned to appropriation and de-Arabisation. Arab-Palestinian food was therefore presented as Israeli, biblical or as part of the Mizrahi-Jewish food culture, and thus belonging to the new Jewish-Israeli state, national identity and culture.

Additionally, the story of Israel's food culture also demonstrates that romanticisation and imitation were not limited to the images and cultures of the Bedouin and the fallaḥ. As we have shown, from a Jewish-Israeli perspective, the local Arab-Palestinian food culture still contains a reservoir of useful and 'fascinating' elements. Though this process has continued unabated since the settlers' early arrival, it has in recent decades taken on new forms, which is to be expected from the changing political and demographic realities.

The modern process is based, as before, on incorporation leading to appropriation, but includes also, at times, the recognition and even celebration of Arab-Palestinian food, including chefs, cookbooks and restaurants. Whether this new 'development' harbours any possibilities for a new cultural dialogue or state of affairs between Jews and Arabs in Palestine/Israel, or at least an opportunity for a shift in relations, remains unclear. From our vantage point, however, and from what we were told by many of our interviewees, the recognition and celebration of Arab-Palestinian food is still limited to particular segments of the Jewish-Israeli population, specifically the dwindling supporters of coexistence. Other interviewees talked about the non-political nature of food and of its 'unique' place, and as such it should not be used to make sweeping generalisations regarding other cultural processes.

Despite the possible optimistic nature of the current process, it is also undeniable that this process is still based on Jewish-Israeli hegemonic and political dominance, as well as the active marginalisation of Arab-Palestinians in Israel, the West Bank and the Gaza Strip, and the denial of their national identity and political aspirations. Our conclusion, therefore, is that the 'celebration' of Arab-Palestinian food in Israel does not seem to challenge the Israeli discourse vis-à-vis the Arab world and culture, but to bolster its boundaries.

Chapter 4

The Creation of an Arab non-Arab Culture:
Between the Arab Other and the Israeli Self

It Ain't Europe Here

Madam Rothschild, did you have enough of Tel Aviv?
Hilton was too full for you, so you went off to Paris?
But you came back home with a scent of Chanel
And went straight back to Ben Yehuda Street
Miss Hipster, well, how is it in Berlin?
Did you fit in easily with the Germans?
Remember that if you had enough with left- and right-wing debates here
You are doing no favours to us – *Gute Nacht und Auf Wiedersehen*

Chorus:
Hop hop hop hopa, here it ain't Europe
Here it's Israel – you'd better quickly fit in well
Hey hey, kapara, here it's not Europe
Here it's Balagan, Old Middle East not gone

Everywhere you go, you feel like Miss Universe
And you work hard in Europe to assimilate with non-Jews
You even have a start-up thanks to your Jewish brain
And you still yearn for a Spanish passport
But remember that in the beginning we're created wild
As Americans with an Arab pride
So, remember, it'll do you no good to loathe
Admit it, you're addicted – to Israel!

Chorus:
Hop hop hop hopa, here it isn't Europe
Here it's Israel – you'd better quickly fit in well
Hey hey, kapara, here it's not Europe
Here it's Balagan, Old Middle East not gone

Hey, put your hands up in the air!
I can easily recognise you from afar
As deep in your heart you are still a child of God

Remember, you're not from London or from Amsterdam
Your wedge, my dear, is from Bat Yam!

Chorus:
Hop hop hop hopa, here it isn't Europe
Here it's Israel – you'd better quickly fit in well
Hey hey, kapara, here it's not Europe
Here it's Balagan, Old Middle East not gone[1]

In October 2014, 'It Ain't Europe Here' became a hit song amid a growing wave of Jewish-Israeli emigration to Europe and the USA. Berlin became the symbol of young Israeli emigrants who left Israel after becoming fed up with the high cost of living and the ongoing political difficulties. This trend of emigration – or at least its discussion – became a prominent media story following the social protest in Israel during the summer of 2011, in which hundreds of thousands of Israelis went out to the streets to demonstrate against the high cost of living and other social-economic matters. In 2014, when an Israeli blogger who lived in Berlin advertised the relatively low German price of a popular dessert (called *Milkey* in Israel) that is also sold in Israel, another 'wave' of protests were sparked – though this time only on the internet and social media – again targeting the high cost of living in Israel. In 2014, things came to a lyrical climax when Israelis living in Berlin released a song titled *Berlin Berlin*, which encouraged their peers to leave the country to find easier, cheaper and more pleasant locations to live in.[2]

The 'response', therefore, of 'It Ain't Europe Here' is fascinating considering the sociological and political context in which it was created. Yet it is also reflective of a deeper phenomenon happening within Israeli culture. The song contains a unique combination of a number of elements. It seems to capture a 'true' sense of Israeliness: a pride in the 'success' of the 'true' Israeli to say exactly what he/she thinks/feels in a straightforward manner (called *dugri talk* in Hebrew, which

---

1   The song 'It Ain't Europe Here' (lit. 'Po ze lo Eropa') was released in October 2014 by Jewish-Israeli singer Margalit Tzan'ani (music and lyrics Doron Madeli). The translation of the song to English is ours.

2   For further reading on Israelis who live in Berlin, see: Ofer Aderet [in Hebrew], 'Berlin is Far Away from Being Paradise for Israelis', *Haaretz*, 15 October 2014: http://www.haaretz.co.il/news/education/.premium-1.2459649 (accessed 30 March 2015). See also Sally McGrane, 'So Long Israel, Hello Berlin', *The New Yorker*, 15 May 2014: http://www.newyorker.com/culture/culture-desk/so-long-israel-hello-berlin (accessed: 30 March 2015). On the issue of the *milkey* online-protest, which began in Berlin, see: Boaz Arad [in Hebrew], 'This is the Person Behind the Milkey Protest' *Haaretz*, 15.10.2014: http://www.haaretz.co.il/news/education/.premium-1.2459652 (accessed: 30 March 2015). To watch the *Berlin Berlin* video clip, see: https://www.youtube.com/watch?annotation_id=annotation_3378956361&feature=iv&src_vid=YCZDiTiCjdU&v=_h7TR0RLBak (accessed: 30 March 2015).

comes from the Arabic *dughrī* – meaning 'straight'); a perception of the Middle East region as a 'mess' (in the song it is mentioned as *balagan*, which comes from the Persian *bala khana* – meaning 'an attic storage place'); a simplistic connection to the unassuming Israeli space, which is mentioned in the song in the use of 'your wedge is from Bat Yam' (the city of Bat Yam is used here as a symbol of the 'average', 'down to earth' Israeli urban space, while the word *wedge* comes from the Arabic *wajh* – meaning 'face') and in the reference to Ben Yehouda Street (a busy, urban, Israeli shopping street); lastly, the Israeli sense of affection (seen in the use of the word *kapara*, from the Jewish-Moroccan dialect, which conveys a sense that 'I would do anything for you') as well as the naïve 'addiction' Israelis have to their country. All in all the song represents a mischievous *mixture* that brings together the different elements that make up the concept of Israeliness and the Israeli way of life – 'Jewish brain' and high-tech, Arab pride, American values, being from the region but not of the region, being God's chosen people whilst having a simultaneous attraction and repulsion to life in the Middle East.

The song also reflects other identity issues in Israeli society. It attempts to push forward a Jewish-Mizrahi sense of Israeliness, in response to the Ashkenazi 'hipster' fleeing to Berlin, and demonstrates the recent Mizrahi renaissance in Israel – a proud reaction to the European-Ashkenazi domination in the country and to the oppressed Mizrahi identity. It also combines elements relating to the gay community in Israel, which is evident in the outfits of the performers in the video clip, as well as in the fact that the video was produced for the fourth anniversary of the Israeli gay events production company *Arisa* (comes from the Arabic *harīsa* – the Tunisian hot chilli pepper paste) which, again, hints at its Mizrahi orientation.[3]

Shortly after the song became a phenomenon with over 400,000 hits on YouTube, the Jewish Agency for Israel, the largest Jewish non-profit organisation in the world, decided to translate the song and disseminate it.[4] This means that for the decision-makers at the Jewish Agency in Israel, this video represented something larger than merely a pop-culture Mizrahi 'response' to Ashkenazi domination. It represented a genuine Jewish and Zionist Israeli approach to the state of affairs in Israel, and to the constructed character of Israeliness. In other words, the song represented, for the Jewish Agency, an 'authentic' and 'desired' Israeli attitude and view – on its 'self' and on its 'other'. It sheds light on what Israelis see, or wish to see, as being 'truly' Israeli.

This is our point of departure for this chapter, which analyses the contrasting nature of Israeli culture, or 'way of life'. It is one which is based on internal dissonances and opposing characteristics; a culture that is in constant negotiation with Arabness, Orientalism and Occidentalism; a culture which wishes to be in

---

3   To watch the official video clip, see: https://www.youtube.com/watch?v=OFZmcSVHnxs (accessed 30 March 2015).

4   To watch the official video of the clip that includes the Jewish Agency's logo and translation to English, see: https://www.youtube.com/watch?v=YCZDiTiCjdU (accessed: 30 March 2015).

the Orient but not Oriental; to be from the Middle East but not of the Middle East; to be Western but not a 'regular' part of the West. These relationships, we argue, are the end result of processes that began before the state was established and while the specificities of these processes have changed over time, discordant elements of the Israeli identity have been endemic since the beginning. To study this phenomenon, we look at a number of selected components of Israeli culture: from paintings to plays, garments to footwear, and from songs to dance steps. The story of Israeli culture, we argue, is one that – like its language, food and national symbols – is in constant conflict between what it wishes to be and what it wishes to negate.

The Complexity of Studying and Analysing Culture

Our main interest and focus in this book has been Israeli national identity and the role Arab and Arab-Palestinian cultures and identities have played in its creation and maintenance. However, our discussion of identity means that we cannot avoid entering into the realm of culture and cultural studies. This is an area that we fully acknowledge is problematic to define: 'discussions about culture have been bedevilled by an inability for theorists to agree on a common definition, for it has remained a fluid term' (Edensor, 2002: 12). Because our focus is on Israeli national identity and the symbols, experiences and actions that shape it, we think it is more useful for our purposes to view culture, in line with Raymond Williams (1983), as a way of life that encompasses and connects a large number of elements and phenomena, for example, values, organisations, practices and products (Hearn, 2006: 208). In other words, our focus lies in human experience, action and interaction more than in high culture or culture with a capital C.

Within the study of nationalism and national identity, culture plays an important part, though the exact relationship between them is a source of contention and debate. The debate over the importance of culture ranges from viewing the nation as a cultural product constructed in response to modernity (Gellner, 1997), to the nation seen as a reinterpretation of primordial ethnic cultural heritage and ties. For us, the modernity argument and the events related to Jewish life in Europe do provide a strong impetus for the Zionist drive to create a Jewish nation-state in Palestine. However, we also agree with Anthony Smith (1991) that the nation was not created out of thin air. It was indeed based on some earlier concepts and notions of a pre-existing, ethnically Jewish culture or cultures. Nonetheless, in order to transform these culturally varied Jewish groups, who may or may not have had some notion of a pre-existing Jewish ethnicity and culture, into a nation, Zionist elites incorporated invented and 'rediscovered' traditions in their attempt to reshape Jewish identity and emphasise the idea of a Jewish nation as rooted and local to Israel/Palestine.

In this regard, and specific to our case study, we view Hutchinson's (2004) ideas of cultural nationalism as particularly useful, specifically the idea of the creation of the nation as a result of the revival of older traditions in response to the rise of modernity. This revival entailed the rediscovery of the homeland 'as a repository of a unique moral vision and primordial energies' and:

> although ethnic groups typically view themselves as linked "ancestrally" to the land, romanticism intensified, extended, diffused and embedded a sense of belonging to a homeland, larger in scale than before, and it proclaimed the defence and the regaining of the national territory as a sacred duty. (2004:113)

The revival of the 'nation' also meant the creation of a new 'unified way of life' that brought together the various 'older' fragments that were perceived to have been lost, as well as the 'new' invented traditions. The subject of 'invented' traditions is particularly relevant when discussing Jewish-Israeli culture and will be discussed in greater detail below.

The idea of creating a new unified way of life fits in with the idea of the Israeli nation as an imagined Jewish community that perceives itself as sharing similar values and sensibilities, particularly with regard to ethnic and cultural ties to the land it inhabits. The imagined Jewish community and its related practices – whether they are invented, constructed and/or organic – give rise to the ideas of *Israeliness* (lit. *Israeliyut*) and to what we think is crucial for our discussion – the Israeli perception of its 'way of life', sometimes titled in Hebrew as Israeli *hayay* (comes from Hebrew *hayayah*, meaning 'being'). These ideas, which can be translated as 'being Israeli' and as 'the Israeli way of life', connect the various cultural, ethnic and historical elements together to form an (imagined) Jewish-Israeli national community. They also entail how the Jewish-Israeli cultural community presents itself externally and internally, and how it represents others to itself. We focus particularly on the 'representative expressions and experiences' that encapsulate Israeliness and Israeli hayay, and the way in which these notions manifest. We are interested in understanding how Israelis see themselves and their society as unique, and how and in what way these notions are reflected.

Do we argue that Israeli culture is homogeneous? Definitely not. The idea of an Israeli culture is, to begin with, a contested one. Is the notion of Israeliness or an 'Israeli way of life' an empirical one? We also don't think so. What we are interested in is not whether such notions truly exist, but how Jewish-Israelis imagine or perceive them. As a result, we are interested in analysing what Jewish-Israelis consider to be 'Israeli'. We therefore don't argue that a uniform Israeli culture exists, but rather that there are a number of practices, traditions, rituals, norms and types of behaviour that are seen as epitomising, or providing a generalised conception of Israeliness and Israeli hayay by Jewish-Israelis.[5] Furthermore, we

---

5 For example, a number of scholars have pointed out the importance of talking frankly (lit. *dugri* – which derives from Arabic *dughrī*, meaning straight) to Jewish-Israeli identity and its straightforward, self-confident, local and 'sabra' associations. See,

don't argue that all Israelis accept, support and or take part in these behaviours, but that they nevertheless recognise these behaviours as 'Israeli'.[6] In other words, we are interested in what can be considered as 'banal Israeliness' (borrowing Billig's 'banal nationalism'), or in the 'stereotypical' manifestations of Israeliness.

It is worth mentioning that Israeliness and Israeli hayay are not static concepts. Indeed they are flexible terms that have changed and are being reinterpreted on a continual basis. For example, writing on the subject of Israeli culture in 1952, Samuel Koenig noted that the central features of Israeli culture were socialism, secularism and collectivism, which according to him were 'threatened' by the rising immigration from non-European Jewish communities.

These 'fluid' terms, of Israeli hayay and Israeliness, therefore, posed a challenge for us in writing this book. How should we work and engage with them? And from when should we start analysing them? We decided to focus on finding patterns within these terms and analysing them according to their given socio-political context. As mentioned earlier, the idea of culture as a way of life also means that there is a direct link between the ways in which culture is represented and manifested, and social and political interactions (particularly in relation to the concept of power). Therefore, and in a similar fashion to Hall (1996), we relate the ways in which Israeliness and Israeli hayay manifest in daily life directly to the power structures and hegemonic ideals and dynamics present in Jewish-Israeli society.

Another element that has helped us to deal with the above-mentioned question was to focus on the tension between bottom-up and top-down processes, vis-à-vis the shaping of Israeli culture. Regev and Seroussi (2004), and Evan-Zohar (1981), as well as others who have examined the history of Israeli culture, note the emergence of a top-down Zionist constructed and controlled idea of Hebrew culture, which formed and, to some extent, still forms the bedrock of Israeliness. This idea, based on particular notions regarding the role of Europe, Western culture and Ashkenazi dominance, had been central to Israeli culture in the pre-state period and continues to manifest itself in various ways to this day. This notion of Israeli culture has been challenged by, for example, Mizrahi Jews, who may resent this Ashkenazi-centric view of Israeliness. Yet it is still evident, for example, in the distinctions made between what is considered to be

---

for example: Tamar Katriel, *Talking Straight: Dugri Speech in Israeli Sabra Culture* (Cambridge: Cambridge University Press, 1986); Don Ellis and Ifat Maoz (2002) 'Cross-cultural Argument Interactions between Israeli-Jews and Palestinians', *Journal of Applied Communications Research* 30: 181–94.

6   Research conducted into the everyday representations of 'Israeliness' emphasise this point of recognising a Jewish-Israeli archetype and group identity, even if many Jewish-Israelis distance themselves from such notions. See, Rakefet Sela-Sheffy, (2004) 'What Makes One an Israeli? Negotiating Identities in Everyday Representations of "Israeliness"', *Nations and Nationalism* 10(4): 479–97.

'Israeli music', which is in fact mostly associated with Ashkenazi singers and sounds, and 'Mizrahi music', which is still considered, despite several changes, to be at the margins of the Israeli musical canon.[7] This tension, between the constructed top-down idea of Hebrew culture and other interpretations and manifestations of Israeli culture, is partly a result of particular ideological arguments, power dynamics and relations that exist in Israeli society (for example, secular-religious; Ashkenazi-Mizrahi; and Jewish-Arab).

The idea of a homogeneous Hebrew culture, as we explained above, has been challenged by many. When discussing Israeli music, Regev and Seroussi (2004), for example, identify three main categories that encompass and form what they define as Israeli popular music: the institutionally supported *Shirei Erets Yisrael* (lit. the songs of the land of Israel) and its more modern equivalents, which have formed the basis for popular music in Israel; Israeli rock, which was to an extent a reaction to Shirei Erets Yisrael; and *Musika Mizrahit* (lit. Mizrahi music), which was and still is a Mizrahi ethnic reinterpretation of Israeli music.

One element within the broad concepts of Israeliness and Israeli ḥayay that has remained central and constant has been ethnic Jewish exclusivity. Jewish exclusivity, particularly in the case of Musika Mizrahit, has been achieved through a process of de-Arabisation. This process of marginalising and removing the appearance and presence of the 'Arab' from Israeli culture can be seen as central to the idea of Israeliness. In other words, Musika Mizrahit can only become 'truly Israeli' when it sheds its Arab image. In building on Gilroy's (2002) study of ethnicity and racism in British society and culture, it is clear to us that what is considered 'Israeli' and the 'Israeli way of life' is based on ethnic absolutism, exclusion and dominance. Therefore, what is of interest to us when we examine Israeliness and Israeli ḥayay, is understanding the relationship between the Zionist and later Jewish-Israeli desire to become native by fostering a unique and local culture through the use of the Arab 'Other' – which includes fascination with and imitation of Arab and Arab-Palestinian images and cultures – while maintaining hegemonic concepts of an exclusive Jewish state.

The Image of the Arab

We identified that the underlying tension in Israeliness and the Israeli way of life exists because of the need for them to be unique, indigenous and Jewish, while being partly based on Arab and Arab-Palestinian elements.[8] Yet in order to move

---

7   One can also think of Russian Jewish migration to Israel as a challenge to the crystallisation of Israeli culture.

8   It is important for us to be very clear at this point that we do not argue that Israeli culture or for that matter that every Jewish-Israeli cultural practice or experience is entirely or mostly based on Arab and Arab-Palestinian elements, but that these are present in what Jewish-Israelis define as Israeliness and the Israeli way of life.

on to discuss the influence of Arab and Arab-Palestinian cultures on Israeli culture, we think it is important to first examine how Arabs and Arab-Palestinians more specifically have been represented in Israeli cultural products. Particularly taking into account the tension we describe above regarding power, indigeneity and ethnic absolutism and exclusion.

Reviewing Israeli films, works of literature, theatre and art, it is clear that the image of the Arab 'Other' has changed over time, but it is equally evident that several important principles and ideas have remained. Broadly speaking, we divide the works into three periods (early Zionist settlement up to 1948 war; the 1950s to the 1970s; and from the 1980s onwards).[9] This division was made based on the specific and unique characteristics we found in each historical episode, as we explain hereafter.

In the early Zionist period, the representation of the Arab 'Other' has been described by Govrin (1989) and Ben-Ezer (1987) as being stuck between two opposing images. Govrin describes the image of the Arab as an enemy or a cousin, while Ben-Ezer talks about the general attitude towards the 'Arab problem' as caught between desires for integration and separation. This duality is partly a reflection of the status of early settlers, a small national minority representing the majority population around them. This duality, at least in the way it has been imagined, did not cease to exist following the creation of Israel in 1948, and instead changed its conscious 'admission' to being Arab-influenced and anti-Arab at the same time. During the 1940s, and especially after the creation of Israel, the policies and mind-sets of separation began to dominate the Jewish society in Palestine/Israel, which created an alleged contrast between Jewishness and Arabness, and between Israeliness and the Arab-Palestinian character. Yet this contrast was needed for the crystallisation of a 'unique' Israeli spirit and characteristic. The Arab influence on Israeli culture 'had' to be suppressed by national needs and desires, and was mentioned only at times when it was needed or when it served an Israeli interest.

Coming back to the first Zionist-European immigrants to Palestine, it was evident on the one hand that there was some level of identification and desire to emulate the local Arab-Palestinians 'in their way of life and manner of speaking, their conduct and their courtesy, the dwellings they lived in, the stories they told, the way they dressed and ate and danced, how they worked their land and herded their sheep' (Govrin, 1989: 15). Locals were seen as something along the line of 'noble savages' with cultural and social traits that were needed by early settlers because the Arab was characterised as 'a true native, deeply rooted in the land and fully at ease in its recalcitrant terrain', in contrast to the 'uprooted

---

9   The question of how to divide these cultural periods has been raised by several authors who have proposed a number of different divisions. For example Ben-Ezer (1997) discusses five important periods with regard to Zionist and later Israeli literature, Bar-Tal and Teichman (2006) and Urian (1997) discuss three important periods with regard to film and theatre respectively.

urbanized Jewish settlers who struggle to overcome their sense of loneliness, alienation and overwhelming inadequacy' (Morahg, 1986: 148–9). The image of the local Arabs as the true natives of the land and role models to be emulated was widely depicted in landscape paintings (Manor, 2002), and as part of the natural scenery of Palestine on stage (Urian, 1997). In particular, early settlers were curious about and fascinated by the images of the Bedouin and the *fallah*, viewing them as a source for Jewish cultural renewal. This tied in with the ideas of a possible common descent or at least some cultural proximity, and with the possibility of coexistence.

On top of this, the Zionist Orientalist perception of both the Bedouin and the fallah – living in tents, herding their sheep, ploughing their fields, and wearing old, unchanged, outfits – captured their imagination. Within cultural production, but also in popular settler imagination, they not only saw them as models, but as ancient *biblical* figures (Manor, 2005),[10] which tied them to the idea of 'return', of Jewish exclusiveness, and of reclaiming the land by imitating its Arab inhabitants.

Yet on the other hand, the local Arab-Palestinians were also represented as a threat and an enemy, replacing the traditional image of the Russian Cossack. Between the Arab enemy and the Zionist settler was a 'deep-seated, never-ending dispute over the same land' (Govrin, 1989: 16). On the stage and on screen, the Arab enemy was represented as primitive, violent, backward, greedy and dangerous (Bar-Tal and Teichman, 2006; Urian, 2005).

One thing that is in common to these two images – of the Arab as a threat and as a model – was the depth of the Arab characters with which Zionist cultural products dealt. They were mostly used as objects upon whom Jewish 'existential fear and anguish can be blamed' (Govrin, 1989: 18). They were represented as highly stereotypical, mostly consisting of three typologies: the fallah, the Bedouin and the *Effendi* (Oppenhimer, 1999; Urian, 1997).[11] We believe that this unholy trinity 'answered' three different existential needs, fears and desires of the Zionist movement. The fallah helped to create an imagined continuity between biblical times and the Jewish 'return'. The Bedouin represented a role model to the new Jew: assertive, independent, pure, simple and brave. And the Effendi represented the threat: a greedy figure reminiscent of the European situation Jews were escaping from, and a figure with whom Jews did not wish to be associated.

As argued by Morahg (1986), among others, early Zionists had an Orientalist interpretation and perspective towards the local Arab population,

---

10   Manor (2005), for example, discusses and analyses Nachum Gutman's painting from the beginning of the twentieth century, and the way in which the image of the Arab-Palestinian fallahi woman with a jar of water on her head was imagined as the biblical figure of Rebecca, while the Arab-Palestinian Bedouin man in his tent was imagined as the biblical figure of Abraham.

11   Effendi was used in the Ottoman Empire as a title of respect for men of high social standing, at times also with regard to government officials and land owners.

which consisted of both romanticisation and objectification. For example, many early artists were interested in the image of the local Arab, as an antithesis to the diaspora Jew, but by focusing on particular images and notions they also essentialised them (Manor, 2002). Interestingly, there is hardly any reference to the 'regular' urban Arab-Palestinian figure, in the main cities – such as Jaffa, Jerusalem, Ramla, Lydda, Nazareth, Haifa and Acre. Urban Arabs were significantly less desired than the 'unchanged' and romanticised Orient represented by the Bedouin, the fallah and the Effendi.

The image of the Arab did change over time, but this mostly reflected the changing values and nature of the Zionist and later Jewish-Israeli society (Morahg, 1986). As we discussed in previous chapters, it is clear that as the conflict between settlers and local Arab-Palestinians intensified, the image of the Arab shifted more towards the 'enemy' and to ideas of separation. This was the case in the context of growing physical and political separation between the Jews and the Arab-Palestinians in the country, which were, all in all, very significant following the inter-communal fighting in 1929 (Cohen, 2013). It was then that a zero-sum game, with regard to 'being Jewish' and 'being Arab' arguably started, contributing to the way Arabs and Jews were represented as separate, and at times, contrasting images.

However, it is important to note that, with the exception of landscape painting, where early artists were particularly interested in the image of the local Arab and his/her environment, Zionist culture focused primarily on the settlers and their struggles, with the Arabs appearing only anecdotally or in the background. In theatre, out of around 100 plays that were put on up till 1948, only 10 had Arab characters (Bar-Tal and Teichman, 2002). In film the Arabs existed mostly in the background and their 'presence as a national entity is completely ignored' (Bar-Tal and Teichman, 2002: 202).

The 1950s to 1970s

After the watershed events of 1948 there was a noticeable change in how the 'Arab' was represented in Jewish-Israeli society. As argued by Ben-Ezer, 'the War of Independence in 1948 marked the transition from an open *Erets Yisraeli* to a closed and besieged State of Israel'. Ideas regarding integration and coexistence, in Ben-Ezer's view, the Arab neighbour, the orange groves and romantic landscapes, all changed their 'status' (1989: 24–5). In place of these ideas, Arabs were mostly depicted as violent, aggressive and an existential threat, and were therefore dealt with in the context of the Jewish-Israeli siege mentality (Bar-Tal and Teichman, 2006). This was echoed also in the Israeli films of the period, in which we identify a process of associating the Arab-Palestinians with terror, blood, guns, kidnapping and violence. Perry argues that the image of the 'Arab' represented the 'instinctive, unknown, dark, sinister, lustful, dangerous, uncontrollable, and unpredictable alter ego of the Jewish hero' (quoted in Bar-Tal and Teichman, 2006: 184). They became the enemy antithesis to the Zionist hero's struggle.

However, even during this period, Jewish-Israeli culture only dealt with the image of the 'Arab' in a superficial manner. According to Morahg (1986: 151), Arabs were again objectified and their character was 'rarely that of a true antagonist' let alone a protagonist. This, we argue, helped to further strip the Arab-Palestinians from being seen as an ethnic and national community – that is, of being a Palestinian *people* – and instead pushed forward their perception as individuals. It was, therefore, a new take on the Zionist myth of Palestine as 'a land without a [Palestinian] *people* for [Jewish] *people* without a land'. Rather than deny the fact of an Arab-Palestinian national presence in the country, this new 'take' viewed them as violent and threatening individuals. By being viewed in this way, the Palestinians 'lost' their peoplehood, and were 'allowed' to exist only as individuals.

Through the process of individualisation and 'foreignisation' of the Arab-Palestinians, there was no longer a need to internalise or be fascinated by 'the Arabs'. Instead, they became voiceless literary devices used to examine Jewish-Israeli identity, values and consciousness (Oppenheimer, 1999). The 'Arab' became a means of symbolising threat and representing the internal anxieties of Jewish-Israeli characters (Morahg, 1986). In theatre, even though still marginal and only inhabiting minor roles, Arab characters were also used to discuss Jewish issues of guilt and the existing socio-economic power dynamics in Israel. Arabs were described as being subservient and as those willing to do the jobs Jewish-Israelis were not (Urian, 2005).

The 1980s and Onwards

In the 1980s, the depiction of the 'Arab' shifted from being seen as marginal and peripheral to Jewish life, to being intertwined with the fate and destiny of Jews in Israel. This was a notion that was obviously not only felt within the cultural field, but in the political field as well. Ami Ayalon, a former chief of the General Security Service (Shabak), has argued that the Jewish and Arab questions cannot be dealt with separately, as they are inherently connected and influence one another. According to him, the Israelis and the Arab-Palestinians are nothing but 'inseparable Siamese twins' (quoted in Klein, 2000), which means that the destiny of the one depends on the other.

The new perception of the 'Arab' also meant that the character became more rounded; 'new Arab characters are no longer static and stereotypical points of moral reference for the central Jewish protagonist' (Morahg, 1996: 151). The Arab character became 'more complex, differentiated, humane, and empathetic', allowing for a more multifaceted view of Arabs and Arab-Palestinians. However, according to Urian it was 'still stereotypic and used to discuss problems that the Jewish society faced' (quoted in Bar-Tal and Teichman, 2006: 201). What is clear is that this 'new' image of the Arab was a product of the Israeli realisation that two people with opposing aspirations were destined to live side by side. Nevertheless, it was still framed within the overarching Jewish-Zionist hegemonic discourse.

This evolution reflected the changing environment in which Jews and Arab-Palestinians existed. Arab-Palestinians, whether living as citizens in Israel or under occupation in the West Bank and the Gaza Strip, became more apparent in Israeli life and therefore provided a point of reference for the dynamics of the Israeli society and state. Examining the relationships between Jews and Arabs, and between Israeli and Arab-Palestinian societies, became a means of exploring tensions and desires (sexual and non-sexual). This was evident in the work of new Jewish-Israeli writers such as Sami Michael, David Grossman and A.B. Yehoshua. However, with regard to Mizrahi-Jewish writers, such as Michael, the examination of 'the Arab' image can also be seen as 'a metaphorical journey of self-discovery for the oriental Jews, who examines his complex relationship with the major Israeli culture, dominated by Western Ashkenazi Jews' (Oppenheimer, 1999: 218).

The relationship between the two peoples and the tension between the various images of the Arab were also exposed and engaged with in Israeli cinema and theatre. Several of the plays that broke new ground with regard to the Arab character were based on books by writers such as Sami Michael, most famously the staging of his award-winning novel *A Trumpet in the Wadi* (1988). This new phase was evident in the cinematic field as well. For example, Rafi Bukai's *Avanti Popolo* (1986) was among the first to critically engage with the representation of Arabs and the mythical Sabra. The film was also among the first to humanise the Arab 'enemy' and to attempt to provide an Arab perspective to the ongoing Israeli-Arab conflict. Nevertheless, as Oppenheimer (1999) argues, it is not clear to what extent Israeli writers 'recognize the Arab's independence and otherness: to what extent does it allow him to have a separate identity, which is not subservient to the Zionist one and to its accepted scheme of values?'

In this respect, it seems evident that Jewish-Israelis struggle with presenting the Arab viewpoint and, by and large, leave this mission for Arab-Palestinians writers and artists (citizens of Israel) who have in recent decades entered the Israeli cultural realm (Urian, 1997). The more rounded image and the new and critical engagement with the 'Arab' image are also the result of a number of developments, among them the various peace initiatives that took place from 1979 (which meant Arab and Arab-Palestinians were viewed as less threatening); an increasing introspective examination of Israel's policies towards the occupation of the West Bank and the Gaza Strip; and the impact of the two Palestinian Intifadas.

Looking at the way in which the image of the 'Arab' has changed over time we believe that three important elements have remained constant and mirror the tension we discussed earlier with regard to Israeli culture. First, Zionist and later Jewish-Israeli writers and artists have continued to exhibit a duality and ambivalence in their relationship with the 'Arab' image/character; a duality that is based on attraction and rejection, admiration and fear, horror and desire, and which corresponds to the earlier Zionist fascination with and disdain towards Arab-Palestinians (Govrin, 1989: 18). Second, despite the more complex depiction of Arabs in Israeli culture, and the fact that they are given a voice and are seen as a people with political aspirations, the engagement with them is still one based on Jewish-Zionist hegemony. In other words, the Arab is still viewed as inferior, associated with fear and terror – and, on top of this – the power dynamic that exists between Jews and Arabs maintain the superiority of the former over the latter (Urian, 1997). This discourse also necessitates the construction and maintenance of the Jewish-Israeli identity in relation to the Arab 'other'. Third, the Arab and the Jewish characters are still presented as mostly separate identities belonging to people from different societies, with an alleged inherent difference. Very rarely are these identities allowed to merge or coexist.

Sing Along: Examples from the Fields of Song and Dance

The above section illustrated the dynamics behind the duality, power relationships and othering that exist with regard to the depiction of Arabs and Arab-Palestinians in Israeli culture. This section will continue to explore these dynamics in greater depth through a number of case studies, beginning with Israeli folk music and dance. We are aware that these limited examples barely scratch the surface of this incredibly large field, yet within the scope of this section, we will shed light on a few examples that tell the story of a unique and somewhat unrecognised phenomenon.

In analysing Israeli folk song and dance, we argue that the focus needs to be not only on what Israelis do in Israel, but also on how they export Israeli culture abroad. We therefore believe that by looking at Israeli expat communities – as part of many other studies that analyse the sociological phenomena of expat communities in the world – we can bring to the fore an essential 'Israeliness'. One can argue that it is often among expat communities that the stereotypical, archetypal or common notions and perceptions regarding national identity can be discovered. The desire to hang on to these perhaps less nuanced elements of national identity clearly demarcates expat communities from their new cultural surroundings, but also highlights what they consider to be essentially Israeli. It is therefore important to examine what Israeli expats consider and define as Israeli.

Interestingly, in many such Israeli expat communities, Israeli ḥayay events or occasions stand out as moments of engaging with and expressing Israeliness. There are a number of practices that routinely take place in such events, in which falafel, pita and hummus will almost always be served. But perhaps none define and resonate with their Israeliness more than public singing (*shira betsibur*) of Israeli folk songs (*Shirei Erets Yisrael*) and folk dancing (*rikudei 'am*). In many ways, these practices best encapsulate the stereotypical and mythical image of the Israeli Sabra. Furthermore, and even though we are discussing the stereotypical Israeli image, this does not mean it stays in one's imagination only.

As a matter of fact, folk dancing and public singing are still prominent features of Israeli society today. It is estimated that tens of thousands of Jewish-Israelis practise folk dancing on a regular basis (Lee, 2011).[12]

The romantic ideas of cultural nationalism, which we discussed above, place an emphasis on reviving, and at times inventing and rediscovering, cultural practices and folk 'traditions'. These are seen as part and parcel of unifying and uniting the nation, giving it a sense of belonging, and rooting it to a particular land, history and narrative. The idea of revival was indeed central to the Zionist movement. However, in contrast to other romantic nationalist movements, Zionism mostly rejected its (European) past in search of a more ancient, Jewish, localised past, which was at times unknown to them. This means that the subject of 'invented' and/or 'rediscovered' traditions is particularly relevant when discussing Jewish-Israeli folk.

Israeli folk music (*Shirei Erets Yisrael*) and folk dance (*rikudei 'am*) are perhaps the two traditions that most exemplify Israeliness in the modern Israeli consciousness. The two will be forever ingrained in the nation's memory because on 29 November 1947, after the UN vote on partitioning Palestine into a Jewish and an Arab state, the whole country appeared to break out into spontaneous dancing and singing – in which Israeli folk music and folk dances were central. The two are also closely associated because many of the songs and dance steps were thought of either as complementary or composed/choreographed specifically to go with one another. Additionally, these two folk practices have been used by the state to further the Zionist cause, through their use in national ceremonies and religious and public events. As a result they received institutional support, and were promoted by the school system, the kibbutzim movements, the Histadrut (the general organisation of workers in Israel), the Israeli Defence Forces (IDF), political parties and others.

One of the problems of examining Israeli folk dance and music is that we tend to view folk traditions and practices as being based on bottom-up, organic and cumulative processes that develop over time in a specific area and that represent the heritage and traditions of a particular group of people. According to one of the founding mothers of Israeli folk dancing, Gurit Kadman (1952), because of the lack of unifying and common practices among Jewish immigrants in Palestine, there was a need to invent folk traditions. As a result, what is considered Israeli folk is actually constructed and relatively new. Two important issues arise from the absence of historical folk traditions and their reinvention: where did the settlers draw their inspiration to construct these folk traditions from, and what ideas and values did they hope to represent and express through them?

The early settlers, and later on the Jewish-Israeli cultural and national entrepreneurs and their institutional supporters, drew their inspiration from a diverse range of elements. Regev notes how 'shirei erets Israel' were based on

---

12 We will come back later in this section to the popularity of Israeli folk practices.

the incorporation of 'oriental musical elements into' the settlers' 'essentially East-European dispositions' (Regev, 2000: 230). This combination between East and West is evident and important in Israeli folk culture. Roginsky, discussing folk dancing explains that 'it was an "invented" tradition of European Jews, motivated by romantic and orientalist ideology, who used the traditional folklore of Oriental Jews and local Arabs in order to legitimize the creation of a modern national culture' (2006: 248). In this regard, it is clear that the musical and dance influences that inspired early settlers encompassed a wide range of elements. In terms of dance, the settlers incorporated and performed the *hora* (Romanian), the *polka* (Polish), the *rondo* (Russian), the *dabkeh* (Arab-Palestinian) and the *cherkessia* (Circassian) among many other dances.

According to Roginsky (2007), there are six different discernible groups that helped shape Israeli folk dancing: the 'old yishuv' Jewish population; Hasidic Jews; Yemenite Jews; Eastern Europeans Jews; Jewish-Mizrahim; and the local Arab-Palestinians, including Muslim and Christian Arabs, as well as the Druze and Circassian populations. Of particular importance were the Yemenite and local Arab-Palestinian traditions because these were viewed as closer to ancient Jewish traditions and thus essential for the creation of 'new' old biblical dances. Kadman (1952) explained that early Jewish sources indicate a great richness and variety of dances used for a large number of occasions. However, most of these dances were 'lost' and, therefore, needed to be rediscovered and or invented. This is where the inclusion of local and Yemenite folk dances appears important. Unlike the process of de-Arabisation that occurred with food and other elements, there was no corresponding process with regard to Israeli folk dancing; though it is important to note that the prevailing power dynamics were very clear, a point to which we shall return to below.

In a similar fashion, Israeli folk music was also a product of a wide range of diverse and eclectic Eastern and Western influences. Nevertheless, Eliram (2006) notes the importance of local Arab-Palestinian sounds, melodies and musical instruments on Jewish musicians in Palestine/Israel, for example, prominent and award-winning prolific composers and songwriters such as Gil Aldema, Yedidya Admon and Emanual Zamir. Regev, however, cautions that the local Arab-Palestinian music did not have a profound impact on Israeli music, but that it served more as general inspiration.[13] This mostly meant adopting popular 'Arabic rhythms and melody structures' (Regev, 1995: 440). The local Arab elements, combined with musical ideas brought from Europe and Hebrew lyrics, produced a musical fusion that became Israeli folk music, but, as Regev,

---

13   This point was made very eloquently by poet and columnist Yehuda Karni in 1922: 'you must keep quiet for a while, and in your silence try to free yourself of all impressions and sounds in which you had been engulfed in the Diaspora ... the wild cry one hears in this country at an Arab wedding is more significant for the Hebrew art of the future than the formal European tune ...' (quoted in Hirshberg, 1995: 254).

explains this does not mean that Israeli folk music sounded anything like local Arabic music.

'Shirei erets Israel' and 'rikudei 'am'' were used as practices to express the ideals and aspirations of the new and growing Jewish-settler community. They often contained biblical motifs, either of the return to the land or of particular stories, though they attempted to pour new meanings into these while taking them out of their traditional religious context (Eliram, 2006). The songs and dances were also used as a way of celebrating the relationship between the community, agricultural work, and the new-found social structure and Hebrew identity. As such they were promoted as acceptable forms of social and cultural practice. Through communal dancing and singing, the folk traditions came to symbolise and celebrate the new social structures and aspirations of the Zionist settler society, as well as its distinctions, sense of belonging and struggles (Regev, 2000; Roginsky 2006).

The success of these folk practices owed much to the institutional support they received. Folk dancing was celebrated and promoted through festivals, such as the one in Kibbutz Dalia and later in the city of Karmiel, but also through its inclusion in national events and ceremonies created by the state (Torstrick, 2004). Leading state institutions used folk dancing as a way of conveying and performing new national ideas as well as celebrating Jewish holidays such as 'Tu Bishvat' and 'Shavu'ot' in a secular manner. Additionally, folk dancing and public singing were also used by the state in its efforts to create a homogeneous Jewish-Hebrew national culture. They were promoted through official bands and troupes supported by the Histadrut and the army (the military bands, lit. '*ha-lehakot ha-tseva 'iyot*') as well as through their inclusion in the national education curriculum and in school performances (Regev, 2000; Roginsky, 2006).

Though the types of songs and dances described here were prominent in the pre-state period and in the first few decades after Israel's independence, they still hold an important place in what Israelis define as Israeliness and are associated with the Israeli way of life. According to Eliram (2006), Lee (2011) and Regev (2000), folk dancing and singing have been making a comeback in recent years. Regev suggests this might be happening in the context of a desire among some segments of the Jewish population to return to their roots as a response to the impact of globalisation. In this regard it is interesting to note that, over the past two decades, public singing and folk dancing have also returned to the national stage and have been more prominently discussed and broadcast on national TV. Eliram (2006) mentions the importance of public singing (*shira betsibur*) of Israeli folk songs as a mechanism of stressing belonging and togetherness in difficult times and its performances at times of national distress, such as recent wars.[14]

---

14   In this regard, Eliram provides a number of useful quotes from prominent Jewish-Israeli writers and broadcasters regarding this phenomenon. For example, journalist and broadcaster Yaron London wrote in 2002: 'here, almost as a surprise, public singing has returned to the centre of the Israeli experience. Singing on television, singing in the periphery and singing in the cities, in the regions of beer connoisseurs and grass smokers.

What is most remarkable about 'shirei erets Israel' and 'rikudei 'am', other than their national importance, is that while they are based, among other things, on local Arab-Palestinian culture, they are used primarily to distinguish Jewish-Israeli society and identity from the Arab-Palestinians. When we asked about the importance of Arab-Palestinian music in Israel today, Motti Regev told us that one can mostly talk about its absence: it is rarely played on Israeli-Hebrew radio stations or performed on television, and there are few significant Jewish-Arab musical collaborations. With regard to Musika Mizrahit – even though it has included and incorporated some Arab elements, and several prominent Mizrahi singers have sung in Arabic – it has mostly managed to enter the Israeli mainstream through the rejection of its Arabness.

The same relation also exists in dance. While Israeli folk dancing incorporated the Arab-Palestinian *dabkeh* dance, which now has several Jewish-Israeli versions, as well as Druze and Circcasian dances, these were marginalised when it came to public and national dance performances in Israel (Roginsky, 2006). Local Arab-Palestinian dance troupes have been given space to perform in national festivals and events, but only under titles such as 'ethnic dances', reflecting their non-Jewish status and the fact that they are not seen as representative of the state. In this regard, and taking into account the importance of singing and dancing events at all levels in Israel, it is clear that Arab-Palestinian presence has been minimal and rare. These events have been staged in order to celebrate a specifically Jewish-Israeli identity and nation – its return and its struggles. Although the dances and songs might have been inspired partly by local Arab-Palestinian culture, they were used to represent values, ideas and narratives that ignored, excluded or marginalised the place of non-Jews in the Jewish state (Handelman, 2004). In short, Jewish-Israelis incorporated local Arab-Palestinian cultural elements as a means to create meaningful attachments to the land and to an ancient and lost Hebrew cultural tradition, while simultaneously working to obscure the origins of these Arabic cultural traditions and the connection between Arab-Palestinians and the land of Palestine/Israel.

From the Scouts to the Palmach and back: The Zionist Youth Movements

Following the last section, it is interesting to mention that Eliram (2006) discusses another important reason for the unique significance of folk dancing and singing in Israeli life and culture, highlighting the fact that they were also

---

How do they sing? Loudly, and as Israelis do, each tries to outdo and sing louder than his friends. Does this phenomenon express a strengthening and renewed bonding of the Israeli society? Is it a secular replacement for religious prayer at times of crises? Does it represent the trumpet of war or a replacement for cries of fear and forgetfulness? Assumption: it represents all of the above' (quoted in Eliram, 2006: 158).

ingrained in the Zionist youth organisations. We believe that the example of the Zionist youth movements is therefore important, not only because they served as 'greenhouses' for the creation and practice of, and engagement with Israeliness, but also because they provided a space in which, for example, folk dancing and folk singing were developed.

In many communities around the world youth movements play an important part with regard to youth culture and socialisation. They act as intermediate agents and as a link between childhood and the adult world. The role played by the Jewish and Zionist youth movements in Palestine before 1948 and in Israel since 1948 encapsulates this traditional element, but also goes far beyond it. In a similar way to the Palmach, which we will touch upon below, the Zionist youth movements in Palestine/Israel acted as pioneering civil-military groups that sought to epitomise the Zionist ethos.

According to Chaim Adler (2008), most of the Zionist youth movements originated in the European diaspora and were established by Zionist political parties – to which they owed their allegiance. The movements were established in order to teach young European Jews 'Jewish and national values' that included skills related for possible future immigration and settling in Palestine, as well as cultural-oriented themes, such as Hebrew language studies and Zionist values. Even though they were inspired by European youth movements and many had a strong socialist identity, the national aspiration to immigrate to Palestine was at the core of these movements, which pushed forward the idea of the return to Palestine and the negation of the diasporic Jewish identity. In Palestine, the Zionist youth movements played an important role in the process of settling the land and, for example, many of the collective agricultural settlements were established by them. They also helped with the mission of supporting the absorption of new immigrants and with national education for *ha-medina she-ba-derekh* (lit. 'the-state-in-the-making').

In this regard, the Zionist youth movements were also related to and used as vehicles to prepare Jewish youth for military service. This was evident in relation to the Palmach in the pre-state period, and later with the IDF. Almog (2000: 231) highlights this connection between the Palmach, the youth movements and the pioneering 'spirit' that was emphasised in the youth movements. In other words, the 'spirit' was the non-material value 'injected' in the youth movements that was later materially 'translated' to service in the Jewish paramilitary and military forces. Following on from this, we believe that studying the emergence of Israeli culture and of 'Israeliness' has to take into consideration the fluid dissemination of knowledge and practices between the civil and the military spheres, represented by the Zionist youth movements and the Palmach.

A good example for this diffusion and close relations – between the Zionist youth movements and the Palmach – is the Hebrew Scouts movement. Based on a British concept developed by Baden Powell, the first Jewish branch was established by Jewish-German educator Arich Croch in the Hebrew Reali School in Haifa in 1925 and was titled 'The Carmel Wanderers' (Halperin, 1970: 278–9).

According to Tzaban (2008) the Hebrew Scouts refused to join the international scout movement and instead established a separate Jewish one, which was later exported to other Jewish diaspora groups, and over the years has become the biggest and most popular youth movement in Israel. The reason for the refusal to be part of the international scout movement had to do with a number of factors, such as opposition to the gender segregation practised by the international movement, in contrast to the egalitarian aspirations of the 'new Jewish' culture in Palestine, but also more politically driven aspects and related to a Jewish-Zionist refusal to join the Arab-Palestinian scout movement which was established in the country in 1912 (lit. *ittiḥād al-kashāfa al-falasṭīnī)*. The Zionist youth movements did not want to become a part of a Jewish-Arab scout movement – which would have been necessary as part of the international movement – and which would have forced them to give up on their desire to instil Zionist and Jewish values and ideals.

This separation was not coincidental but an early sign of future communal relations. Indeed, two decades after its establishment, the majority of the Hebrew scouters – but gradually also members of other Jewish youth movements – were recruited into the newly created Palmach paramilitary organisation (Weitz, 2000). The process of localisation through activity in the youth movement was therefore 'completed' in the Palmach. As such, special Palmach training units were established titled *Ha-Hakhsharot* (lit. 'The Trainings') in which a group of youth – usually from one of the youth movements – was recruited as a whole (meaning: the whole class or the whole year or the whole age group) to form a separate unit in the Palmach (Almog, 2000: 231). When taking into consideration that members of the Palmach both admired the local Arab-Palestinians and fought against them, it is striking to note that similar dissonances slipped into the Jewish society as a whole, with the youth movements – including their training and education – playing their role in the dissemination of national culture and knowledge. This, we argue, is a process that lies at the heart of the creation of 'Israeli culture' and is one that could not have been achieved without the youth movements and the reproduction of knowledge carried out by them. This reproduction of knowledge was part and parcel of the traffic between the different-yet-similar civil and military spheres in the Jewish-Zionist society. This is evident in the words of Netiva Ben-Yehouda, who is one of the most famous writers about the Palmach, and a former a commander: 'the only thing that kept me in the Scouts was the fact that at the time in order to recruit to the *Palmach* one had to be a member of a youth movement' (Tessler, 2007: 26).

Following recruitment to the Palmach, however, it seems that not only that the ambivalent attitude to the Arab 'Other' not disappear, it was instead strengthened. Groups such as the Palmach instilled and created Hebrew culture, which on many occasions pushed forward particular elements based on Arab-Palestinian culture. Netiva Ben-Yehouda's account of the Palmach in the 1940s is an example of the on-going Zionist admiration of the local Arab-Palestinian population, which remained dominant even in times of heightened conflict. This how she remembers the period (1981: 176):

The Arab inhabitants [of Palestine] were from the very beginning, in the view of the youth of the Palmach frightening, intimidating, courageous heroes ... [The Zionists] envied them .... These Jewish youngsters imitated the Arabs in everything .... Even the "chizbats"[15] [from Arabic *kidhb*, lit. "lie"] and the "finjan"[16] they took from the Arabs .... The Arabs represented the role model for the people of the land, and while we were not well defined, we definitely weren't "Diasporic Jews" .... Anyone who could say a few words in Arabic had more prestige, and if he had Arab friends – he was a real king. Even one Bedouin friend ... or a porter in the Jaffa port was enough ... whoever knew more Arab customs, and knew how to behave among them, and how to create a common language with them – was for us more than God.

It is interesting to analyse this quote and to keep in mind that the Palmach was not a cultural organisation but a paramilitary force that was created in order to push forward, protect, enable and maintain the Zionist national project in Palestine. However, such groups also served to facilitate the transfer and incorporation of Arab-Palestinian cultural elements into other social spaces – for example through the youth movements – to the extent that Israeli culture today still celebrates the Palmach, and its love-hate relationship with the Arab 'Other's' image, culture, food and music. Given the 'social' and cultural importance of the Palmach, it is indeed interesting to draw parallels with the Zionist youth movements.

The importance and centrality of the youth movements to the construction and maintenance of Israeli culture is widely accepted. In her study on the importance of fire rituals, Katriel (1987) argues that those who wanted to belong to the new emerging Hebrew and later Israeli identity had to be part of and accept the rule of the youth movements. Many of the practices that later became part and parcel of the Israeli hayay were engaged with and incorporated through the youth movements, and were identical to Arab-oriented habits adapted by members of the Palmach. These included, for example, Oriental sitting, Israeli sense of dress (which we will discuss below), *chizbats*, folk dancing and *Yedi'at ha'arets*,[17] alongside Hebrew-Arabic slang and particular, Arab-oriented food items. In this regard, it is important to note, that many of the songs and dance routines that were later popularised were first practised and learned in the youth

---

15 *Chizbat*, a word adopted from Arabic into Hebrew, is a category of oral story telling tradition that brings together myth, reality and humour. It comes from the Arabic word for lie, '*kidhb*', and is based on the local Arab culture of telling 'tall tales'.

16 Arab small cups of coffee.

17 In this regard, Yael Zerubabel (1997) stressed that the fieldtrips (as part of the *Yedi'at haarets* activities in the scouts) stood at the heart of the Jewish youth movements in Palestine/Israel, and were considered as a 'sacred activity through which Hebrew youth could reclaim their roots in the land' (1997: 121). These fieldtrips were seen as part and parcel of the process of localising, just like the admired image of the Bedouins and the fallahs who knew their land, and its herbs and paths.

movements. As Adler (2008) notes, the youth movements should be seen as social laboratories for the testing, practising and inculcating of Israeli identity and culture, and values that were 'proved' local, were later to become popular in the Jewish community in Palestine and, from 1948, – among the Jews in Israel.

This admiration and fascination of the local Arab-Palestinians, in particular the fallaḥ and the Bedouin, was common among settler-Zionist youth in the pre-state period, and was part of the complex duality that also viewed them with suspicion and as a threat. Yet the existence of the same dissonant view of the Arab in Zionist ideology – from the youth movements to the Palmach – should not be a surprise. After all, the youth movements were related to Zionist political parties and, as such, were part of the political environment in Palestine. This included political protest, opposition to Arab nationalism and support for campaigns such as the one advocating for '*Totseret Ha'arets*' (Dror, 2008).

The importance of the Zionist youth movements is also of course related to their size. As a matter of fact, in the pre-state period, around a quarter of Jewish youth were engaged in the youth movements (Tzaban, 2008). This figure was still very high in the 1950s (Eisenstadt, 1951), but steadily dropped in later decades. Today around 180,000 Israeli youths are part of the Zionist youth movements. Some, such as the scouts' movements (with a uniform that resembles that of the IDF) and Bnei Akiva (especially with regard to religious-Zionism and the settler movement), still play an important part in Israeli culture. Though based on different political orientations and parties, they are all part of the Jewish-Israeli, Zionist 'culture'.

What do Israelis Wear?

As we demonstrated in the previous chapter with regard to food, Jewish-Israeli fascination with, and incorporation of Arab and Arab-Palestinian cultural elements has, despite arguments to the contrary, continued unabated. Though degrees vary, Arab and Arab-Palestinian cultural elements are incorporated constantly into Jewish-Israeli society. The story of Israeli fashion is a good example of the continuation of this phenomenon and its impact on modern daily life. What is interesting about the story of Israeli fashion is that it shows that the initial process of fascination and imitation leads to appropriation at various levels and is carried out by different groups within Jewish-Israeli society. It is also clear that different segments of Jewish-Israeli society adopt and appropriate Arab-Palestinian elements for different reasons and purposes.

Fashion might appear an odd choice through which to examine Israeli society. For decades the fashion image projected by Zionist leaders and later on by the Israeli state was that of moderation, utility and frugality. This was partly motivated by the collective and socialist ideals many Zionist settlers brought with them from Europe, but was also necessitated by the promotion of austerity measures by the Israeli government in the decade after 1948 in order to deal with mass migration and economic development. According to Almog, in terms of fashion, the Jewish-Israeli Sabra was presented in Zionist imagery as wearing 'khaki shorts and blue shirts characteristic of the [Jewish] socialist youth movements, pinafores and rubashka shirts influenced by Eastern European style, Bedouin kūfiyyās, biblical sandals, and dome-shaped caps known as kova tembel' (2015: 19). Even this short description of the desired Sabra image contains fashion elements that were 'borrowed' from the local Arab-Palestinian populations, such as biblical sandals and *kūfiyyā*s, which we dealt with in the second chapter. However, there was a discernible gap between what most Israelis wore or wanted to wear, and this idealised and stereotypical image.

Images are an essential and important part of the world of fashion. The Jewish and later Israeli experience of urban living in Palestine/Israel was very different from the pioneering image of working in collective agricultural settlements promoted by the Zionist leadership in Palestine (from the end of the nineteenth century). In the urban environment, Jewish and later Israeli fashion ideas and modes were influenced more by international trends than by ideas of utility. As a result, Israeli fashion was much more in line with the trends of individuality and Americanisation, as can be seen in the importance of wearing jeans, and which also fitted in well with the Jewish-Israeli ideas of *chutzpah* and '*dugri* talk' – of wearing what you want and when you want.

Fashion also became one of Israel's leading exports, second only to diamonds in the pre-high-tech boom years. However, one may still ask what influenced particular Israelis, Israeli fashion designers and trends? And here the importance of the Arab world and local Arab-Palestinians becomes more apparent. For example, in the 1960s and 1970s, the Israeli fashion industry according to Nurit Bat-Yaar (2010) was fascinated by Oriental images and designs and modelled outfits based on Arab, Arab-Palestinian, Yemenite, and also Mizrahi-Jewish styles and elements, which we believe stems from a desire to 'blur' the boundaries between Mizrahiness and Arabness.

Thinking about Israeli society, you would probably least expect to find Arab-Palestinian elements incorporated by Jewish settlers in the West Bank. According to Almog and Zalsberg (2010a, 2010b, 2010c, 2010d) Jewish religious-Zionist settlers in the West Bank exhibit a number of different approaches to fashion based on age, geography, levels of religiosity, ethnicity, socio-economic status and political affiliation. Interestingly, there are a number of dominant features among settlers, men and women alike, which stand out and have to do with head- and foot-wear, and both of these fashion features are also – very much – Arab-Palestinian. With regard to footwear, among both male and female settlers the custom of wearing sandals is very prominent. We have written about the issue of wearing biblical sandals in the second chapter in detail and we will come back to this point below.

Most religious Jews cover their heads with a skullcap, a tradition that started in the seventeenth century and that has gained prominence over the past century, particularly in Israel/Palestine. Among religious-Zionist settler men, a recent and growing trend has developed of wearing larger skullcaps that cover the entire head. The size of the new skullcap is used at times as an expressive element of faith. Where this tradition has come from is unclear and it is evident among a number of other Jewish-Israeli religious groups. Nevertheless, what is fascinating about this phenomenon is its proximity, both in terms of fashion and geography, to the style of the *ṭāqiyya*, the Muslim skullcap. The ṭāqiyya was usually worn in Palestine only by those who had gone on the *Ḥajj* (the Islamic pilgrimage to Mecca), but has in recent decades gained prominence among those who have decided to 'return' to their faith. While it is impossible for us at this stage to confirm whether the Jewish tradition was/is influenced by the Arab-Palestinian Muslim tradition, it is hard to imagine that Jewish settlers in the West Bank were not aware of the similarities to the Islamic ṭāqiyya. It is important to explain that the modern Jewish-settler inspiration for wearing the larger skullcap in order to demonstrate their faith is not a tradition that was brought to Palestine/Israel by Mizrahi Jews. We would image that this decision was influenced by the settlers' encounter with the Arab-Palestinian Muslim 'Other' in the West Bank. The reason why we think this is an important issue is that it is part of a wider process of adoption and appropriation of Arab-Palestinian fashion items by Jewish settlers. The case of the religious-Zionist settler women makes this point clearer.

In the 1970s a fashion began among settler women of wearing a women-*kūfiyyā* and or a shawl as headwear. According to Almog and Zalsberg (2010a), at times it is hard to tell these apart because of the use of similar material. Interestingly, the shawl, which is an Arab scarf (originally from Persian: *shāl*) that the Zionist immigrants to the country first saw around the necks of Arab-Palestinian women, has since then gone through a long process of de-Arabisation and de-Palestinianisation. For example, when tracing the 'development of dress patterns and external appearance in the kibbutz', Mikha et al. (2009) highlight that, from an early stage, the Zionist immigrants to the country had used them both as items of dress that were imitated from the Arab-Palestinians working in the fields, as well as items used by Arab-Palestinians in dance, especially the *dabkeh* dancing. They write:

> In the folk dancing in the Kibbutzim, a colourful piece of cloth was an important part of the custom. The senior members of the kibbutzim say they still remember Zipora Zaïd (the wife of Alexander Zaïd, one of *Ha-Shomer*'s founders) herself a member in Ha-Shomer, leading a mass hora dance, waving in her hand with a red kerchief ...

Almog and Zalsberg (2010a, 2010b) explain that the wearing of the kūfiyyā/shawl, as well as the biblical sandals and the larger skullcap, are among a number of different fashion trends and statements of religious-Zionist setters, including a 'new age' bohemian style (in Hebrew *zaruḵ* – lit. unkempt). Nevertheless, it is also a reflection of the settlers desire to resemble and emulate their Zionist pioneer predecessors as well as their biblical forefathers and foremothers, be it Abraham and Jacob, Sarah or Rachel. By wearing traditional Arab foot- and head-gear, they seek to connect with their Jewish biblical past, as well as emphasise their indigenousness and rootedness. In so doing, the settler community in Israel is an amazing living example of a process that used to be central in the European-Zionist localisation and imagination upon immigrating to Palestine.

Furthermore, and just as fascinating, one should also take into consideration that one of the dress elements used by settlers who emphasise the 'new age' bohemian style is the *jallābiyya*, the traditional Arab long dress worn by men and women. Even though the jallābiyya was worn by prominent Jewish-Israeli cultural figures, such as Dan Ben-Amotz in the 1970s, as a symbol of Tel Aviv's bohemian life style and as a symbol of localness and Israeliness, and has featured among fashion designers ever since, it is only in recent years that it has risen to prominence. While some have linked its new found fashion status to second and third generation Mizrahi Jews rediscovering, hinting to and celebrating their cultural roots, it is important to clarify that the jallābiyya was not widely worn by Mizrahi men, and was rarely worn by Mizrahi women. In fact, the discovery and 'return' to the jallābiyya can be seen as part of a wider attempt to reconstruct Jewish, Israeli and Mizrahi identities through the appropriation of Arab, Arab-Palestinian and Mizrahi cultural elements (Madar, 2015; Yaakov, 2012).

In 2011 two Jewish-Israeli entrepreneurs opened 'Galabiyya' a boutique fashion shop specialising in 'urban gallabiyyas' in a variety of designs and colours in down town Tel Aviv. They claimed in an interview with *Haaretz* newspaper that, although it is mostly seen as an ethnic dress and many Israelis were uncomfortable with wearing it, they were hoping it would catch on (Atwan, 2012). Interestingly, by 2014, the jallābiyya had become one of the hottest and to a certain extent bestselling fashion items in Israel. At the end of 2014, though, when leading Israeli fashion designers were asked by Israeli Channel 10 to give their views on the previous fashion year, several stated that the jallābiyya trend had lost steam and was already overstated, which just goes to show how quickly things change in the fashion world. The idea behind using the jallābiyya as a fashion item was articulated by one of Israel's leading fashion designers Tamar Primak. In 2013 several Israeli fashion designers were asked to create a collection that paid homage to former Israeli fashion icons. In creating her collection of trendy beachwear jallābiyyas, Primak explains that she was paying homage to Israeli fashion in the 1980s and to Mediterranean swimwear (Katzir, 2013). What is clear from the jallābiyya example is that the end process of appropriation and de-Arabisation leads to a partial denial of its Arbness and Palestinian-ness.

Constructing a Jewish Home? Neo-Orientalism, Architecture and Gentrification

As we have shown throughout the book, the process of imitation and appropriation leading on to localisation and marginalisation through denial manifests itself in many aspects of Israeli identity, culture and society. One particular area where the cultural aspect of localisation and marginalisation is apparent, and which leads directly to the physical and social exclusion of Arab-Palestinians, and at times even to expulsion, is architecture. We argue that Zionist and later Israeli architecture is a prime example of how culture has a direct impact that goes beyond the mere representation of a society's way of life. This example demonstrates the importance of understanding the power relations that existed and still exist between Jews and Arabs in Israel/Palestine and the significant exclusionary drive (physical and cultural de-Arabisation) that lies at the core of Zionist and Jewish-Israeli way of life.

The drive to create a unique Zionist-Hebrew culture in Palestine, which we discussed above, was also present in the way the country was established, vis-à-vis its architecture. Early Zionist architects sought to create a unique Zionist-Hebrew style that would encapsulate the settlers' important European, Jewish and Zionist technical knowledge, ideals and values, but also localise them by ingraining them in the landscape – which was Arab in nature – and establishing a temporality between them and their perceived biblical heritage and past. In order to achieve these aims, prominent Zionist architects, such as Alexander Baerwald, Jospeh Barsky, Eliezer Yalin, Joseph Manor and others, tried to create a synthesis between European and Middle Eastern elements, and between the past (which was exemplified by local Arab-Palestinians) and the present, that would integrate Jewish, Western and Eastern values and know-how. This style, which became known as 'Erets Yisraeli', was part of a wider cultural and art movement known as the 'Erets Yisraeli School', dominated Zionist architecture in the first few decades of the twentieth century. On many occasions when the 'Erets Yisraeli' concept was mentioned, be it 'Erets Yisraeli Arabic' or 'Erets Yisraeli gazelle', the word 'Erets Yisraeli' was synonymous with 'Palestinian' and therefore encapsulated the replacement through imitation the settlers practised.

The 'Erets Yisraeli' style imitated, adopted and utilised manifold Ottoman, Arab and local Arab-Palestinian architectural elements, for example, the use of arches and arcades, open spaces, circular domes, Arab and Ottoman-style façades, small windows, minarets and local stone as prominent building materials. Architects we spoke to said that, beyond the political and Orientalist ideals held by early Zionist architects, one of the main reasons for using Ottoman, Arab and Arab-Palestinian elements was instrumental: why try and solve technical and practical problems if local solutions, based on hundreds of years of experience and knowledge, had already existed. Alexander Baerwald, for example, who exemplified the 'Erets Yisraeeli' style and who was one of the most important Zionist architects in Palestine and the first professor of architecture in the Technion – Palestine's and later Israel's institute of technology – explained that the use of local Oriental elements helped architects to engage and deal with a number of issues: the importance of dealing with and capturing water and light; utilising local building materials, particularly stone; making use of local knowledge and skilled labourers, who were available and inexpensive, but who were less familiar with European styles and techniques; and the need to address local environmental and geographic concerns (Ben-Artzi, 2006; Levin, 1980, 2006).

The synthesis of styles and the importance of Ottoman, Arab and Arab-Palestinian elements can be seen clearly in many of the prominent buildings built by Zionist architects in the first three decades of the twentieth century in Palestine, and which came to represent the Zionist and Hebrew community's achievement and values. These included, for example, the Hebrew University in Jerusalem, the Hebrew Reali school in Haifa (see Figure 4.1), the Hebrew Gymnasium Hertzelia (in Tel Aviv), the Hadassah University complex in Jerusalem, and the Technion in Haifa. Dolve and Gordon highlight, however, that 'many buildings in the early Zionist period were orientalist only in their façade and may be [seen as] caricatures of genuine Arabic architecture' (1992: 363). The term Orientalist is used in this instance by Dolev and Gordon to express the 'use of indigenous Arabic architectural elements in constructing Jewish building'. Though also utilised to some extent by other Zionist/Israeli architectural styles, such as the 'eclectic' that featured prominently during the British Mandate period, the use of Oriental elements slowly decreased in Palestine/Israel as the 'Erets Yisraeli' style became less prominent.

In a similar fashion to some of the examples we discussed above, and specifically with regard to food culture, the use of Oriental architectural elements saw a revival after the June 1967 War, known in Hebrew as the 'Six Day War', and the Israeli re-encounter and 'rediscovery' of the West Bank and East Jerusalem,[18] in particular their Jewish, nationalistic and biblical importance. The importance of June 1967 is particularly relevant in relation to the Israeli control and redevelopment of the Old City in East Jerusalem, which also included the destruction of the Mughrabi neighbourhood and the expulsion of its residents. As a matter of fact, it was the first urban planning decision that was taken after the 1967 war, made a day after the war ended. Jerusalem's then mayor, Teddy Kollek, toured the Old City with Ben-Gurion, and both agreed that the 800-year-old Mughrabi neighbourhood should be demolished in order to create a ceremonial national plaza in front of the Wailing Wall – the holiest site for Jews. Hundreds of Muslim inhabitants were ordered out and their houses were demolished. The Western Wall Plaza became a reality almost overnight, and the Israeli flag was planted on the site where the houses had once stood, symbolising the triangular relationship between the state, Jewish religion and collective forgetfulness in post-1967 Israel. Interestingly, and while the future

---

18 The West Bank, including East Jerusalem, was under Jordanian control from the end of the 1948 war until June 1967.

Figure 4.1   The Hebrew Reali School in Haifa in the 1940s
Source: Ccourtesy of the Archive of the Hebrew Reali School in Haifa.

of the rest of the territories occupied in 1967 remained a matter for debate, there was no doubt about what was to be done with East Jerusalem: it was to be 'unified' with West Jerusalem, at least according to Israeli law, in order for it to become an integral part of the State of Israel (Mendel, 2013c: 39–40).

Kroyanker (2006) mentions that the then Israeli Housing Minister Mordechai Bentov said in 1967 in a speech to the Knesset that the future buildings in the newly unified city of Jerusalem would be Orientalist in nature. Bentov instructed Housing Ministry officials and urban planners to align and integrate the new and redeveloped buildings and neighbourhoods with the local landscape, in particular the Old City and its prominent walls, and to project an image of authenticity that would connect the new Israeli structures and society with Jewish biblical Jerusalem, and, even if it was not said out loud, with the style that dominated the Arab neighbourhoods of the eastern side of the city. In other words, it became official policy to imitate and adapt local architectural elements in order to construct Jewish-only neighbourhoods in Jerusalem that would emphasise the Jewish historical connection to the city, and that would practically enable the 'unification' of the city.

According to Mendel (2013c: 43–4), it was architecture that was given the most far-reaching role in 'uniting' the city. Construction in Jerusalem after 1967 was to forgo the utilitarian modernism that characterised the architecture in Jerusalem over the years 1948–1967. During those years the question was how to build the maximum residential units with minimal expenditure. The solution found over the years 1948–1967 in West Jerusalem – but also in other cities – was a monolithic take on the International Style: rectangular blocks, which gave these new neighbourhoods a rather boring character; square, in both the geometrical and the slang sense.[19] By contrast, the post-1967 construction style responded to what Israeli authorities saw as a new set of problems: their unprecedented sovereignty over the Old City, including the holy sites of Muslims and Christians; international criticisms of Jewish-Israeli neighbourhood-settlements built on land confiscated from Arab-Palestinian villages; and the difficulty of creating a continuum between the western neighbourhoods and those in the east, which were also built on newly confiscated Arab-Palestinian land. Their solutions – simulacra of 'historic' styles and surface 'cladding' would make 'unified Jerusalem' the most postmodern of cities.

Strikingly, the architects selected by Jerusalem's Mayor Teddy Kollek were first sent to the Old City as part of their fieldwork, to soak up ideas and inspiration. Filled with the euphoria of the 1967 military victory, they concluded that a neo-Orientalist style would be most appropriate for an Israelified Jerusalem, demonstrating how aesthetically sensitive the Israelis were to the region's cultural heritage; how naturally they blend into the landscape; and how their present architectural style was similar to the Arab and Oriental style, and indirectly to their own ancient past in the city. Features of an Orientalised-Arab architecture – arches, gates and domes – were adapted for modern construction techniques and became part of the landscape of the 'new Jerusalem'. The style corresponded closely to several key political ideas: the Israeli stress on the 'return' of the Jewish people to their Oriental 'roots'; the need to forge a unification between old (and biblical) Jerusalem and the new housing projects that would downplay the act of occupation; and an extension of the Zionist colonialist paradigm of bringing modernisation and

---

19   See: David Kroyanker [in Hebrew], *Jerusalem: Neighbourhoods and Houses, Periods and Style* (Jerusalem: Keter, 1996), p. 190. Kroyanker, an Israeli architectural historian, sees the buildings in Stern Street (Kiryat Yovel neighbourhood) and Ha-Nurit Street ('Ir Ganim neighbourhood) as the ultimate examples of this style: eight- and nine-storey buildings with no elevator, which were constructed during the 1950s under austerity measures.

development to the 'unchanging East'. In reality, as the Israeli architectural historian Zvi Efrat has argued, this so-called 'contextual' architecture involved formless clusters of 'sentimental buildings, influenced by alleged "regional" connections' – 'pseudo-historical creations of Oriental and Mediterranean mimicry', said to embody 'an association with antiquity and national roots'.[20]

Another important process that brought about the use of 'Oriental' architectural elements has been the gentrification associated with the urban planning of 'mixed cities' in Israel, in particular, Jaffa (Yafo in Hebrew, Yāfā in Arabic), Lydda (Lod/al-Lid), Acre ('Akko/'Akkā), Ramla (al-Ramla), Haifa and Jerusalem (Yerushalayyim/al-Quds). We use inverted commas for the 'mixed cities' not because they are not cities in which Jews and Arabs reside, but for two reasons. First, they represent spaces in which Jewish neighbourhoods/areas exist alongside Arab neighbourhoods/areas. Second, most of these cities were Palestinian cities with an Arab-Palestinian majority before 1948 and, following the *Nakba*, the Palestinians who remained were not there as a result of a 'mixed' Israeli policy but of an unsuccessful attempt to empty the cities of their Palestinian inhabitants (see, for example, Yacobi, 2009).

The new ethno-gentrified neighbourhoods of the 'mixed cities' borrowed Arab and European architectural elements to create a modern sense of localness while replacing, segregating and pricing out the Arab-Palestinian residents. In this way, the new push for 'Oriental features' drew parallels between Arab architectural features and the idea and image of the Arab village, long associated with poor planning and clustering, with localness and authenticity. This embrace and 'legitimization' of the 'Orient' – the Arab village and Arab architecture, and to some extent also the Jewish '*Mizrahiut*' (Orientalism) – also includes a clear rejection of the 'Arabness' of the Orient (and the Mizrahi) and the exclusion of Arab-Palestinians (Yacobi, 2008).

The 'Andromeda Hill' project in Jaffa epitomises the gentrification process we discuss (see Figure 4.2). The project has been based on the establishment of one of Israel's most exclusive and expensive urban developments at the heart of the poorest Arab-Palestinian neighbourhood in the 'mixed city' of Jaffa as part of the gentrification of the city. It is marketed as a modern, and at the same time authentic, gated community (city-within-a-city) that enables its residents to view the ocean from their windows and experience the rebirth of 'ancient Jaffa'. What is particularly relevant for our argument is the fact that the establishment of a mostly Jewish neighbourhood, presented as the epitome of Israeli architecture, planning and style, but based partly on 'Arab' elements, occurs in the midst of an Arab-Palestinian neighbourhood that had suffered decades of neglect and 'disinvestment'. However, instead of attracting investment for the Arab-Palestinian neighbourhood, the municipal and private planners sought to create an

---

20   Zvi Efrat's text in his exhibition, 'The Israeli Project', held in Tel Aviv in October 2000, quoted in Eyal Weizman *Hollow Land: Israel's Architecture of Occupation* (London: Verso, 2007), p. 47.

Figure 4.2   'Andromeda Hill'

old-new neighbourhood that would appeal to new residents, who would be more affluent, and who would help support the 'older-local' residents. While this pretty much sums up most gentrification projects around the world, the Israeli case is different in that the new residents are mostly Jewish and the 'older-local' residents are Arab-Palestinian (Monterescu, 2009).

Epilogue: Arab Culture in Israeli Society Today

Throughout the chapter we have discussed and emphasised the Zionist aim of establishing a Jewish polity/state in Palestine and the creation of a 'new Jew'. From its inception, the Zionist aim was based on exclusion, the creation of a state for Jews and a separate economy, politics and society, all of which went hand in hand with the creation of a new Jewish identity and culture. The new culture and identity were based on negating the diaspora, localisation, connecting with the biblical past and on elements of romantic nationalism and socialism. This made many Arab symbols a 'natural', non-problematic, part of Israeli culture. The popular *nargilah* for example (water-pipe) is one mere example of that process, which joins many others mentioned in the chapter.

The Zionist aim was based on a double return, a physical return to Palestine/Israel but also a spiritual and ideological return to the Bible as well as their perceived Jewish roots and history. This second return and the importance of bridging the gap of two millennia, necessitated the construction of an eclectic culture based on a wide range of elements, including Arab-Palestinian, Oriental, Mizrahi and even biblical ones. As a result, many of the locals – Jewish, Muslims, Christians and Druze alike – were imagined as preservers of biblical traditions. The cultures of the local population were therefore used in order to recreate 'authentic' Jewish folk traditions that connected the new settlers with their imagined biblical forefathers and mothers.

Thus, in their attempt to become native, and because of the importance attached to indigenousness, local cultural elements were imitated and adapted to become Israeli. As a result, what is considered *Israeli* and the Israeli way of life is based to some extent on local cultural elements. However, and due to the social and political setting – that is, Israel as a nation-state that sees itself as located in the Middle East, but does not see itself as of the Middle East – the importance of indigenousness and exclusion also meant that this influence had to be denied and forgotten and at times marginalised. This process was successful to such an extent that today many of the cultural items and traditions that were appropriated are imagined and perceived by Jewish-Israelis as being Israeli through and through. Additionally, what comes out of the process of creating Israeli culture is that, while actively imitating and adopting elements, these same elements are later used to marginalise and replace local Arab-Palestinians.

Margalit Tzan'ani, from the first paragraph of this chapter, talks in her song about being in the Middle East and not in Europe. Israeli society is indeed changing, and the Mizrahi renaissance Tzan'ani talks about is an example. Yet it is still based on a number of elements, some of them Arab and local, and which can be seen as Middle Eastern, but many of which are read and contextualised within the boundaries of the Ashkenazi, European and Zionist discourse. Therefore, we can find Arab elements in Tzan'ani's song, but as part of a greater context that denies genuine Israeli integration into the region, which altogether keeps the Arabisation potential futile. These, we argue, are the boundaries of the Jewish-Israeli discourse, which encircle the horizons of expectation of all Israelis, including those of Tsan'ani.

Conclusion

A number of years ago, we were invited to a mostly Israeli expat party at a mutual friend's house in north London. All in all, it was a pleasant and enjoyable evening. As most of the guests started leaving, some of us congregated in the back garden for a few last drinks, and some for a cigarette or two. At this stage, and not surprisingly for Israeli-oriented parties, the conversation drifted towards politics.

We don't remember exactly how it came down to it, but at some point the talk revolved around the Arab-Israeli conflict and Jewish-Israeli and Arab-Palestinian relations. In particular, it dealt with whether these two people have much or anything in common? It quickly became clear to us that we were in a minority on this issue. Both of us thought that Jewish-Israelis and Arab-Palestinians have much in common. It simply looked like common sense to us, almost an obvious and instinctive thought, a given that needed no research: if people live in close proximity to each other and have had a shared, albeit violent and difficult history, they would have something, or much more than this, in common. We thought that there would naturally have been some level of cultural diffusion between the two people, especially if they lived next to each other and have experienced changing power relations: from the time that Jews were a minority population in historical Palestine to the current situation in which Palestinians are a minority population in modern Israel.

One of our friends, though, made a statement that startled us. He was a third generation Iraqi-Jew, whose family immigrated to Israel in the 1950s, who then, after finishing his army service, went on to London to study, and later got married and settled there. Yet, according to him, he felt very much 'Israeli' and as such claimed that he, as well as other Jewish-Israelis, had nothing in common with Arab-Palestinians. His main explanation was his conviction that Jews in Israel, be they of Mizrahi or Ashkenazi descent, have created a 'melting pot' society, which is Jewish in its essence, and which had nothing, or very little, in common with Arabs. He was almost mad, or insulted, by our juxtaposition of Palestinian cultural markers and Israeli ones, as for him the two represented complete opposites.

In our view, this was a rather problematic statement to make, and we thought it was obviously related to how he viewed the Israeli-Palestinian conflict. Furthermore, it seemed problematic to us because the 'melting pot' theory in Israel does not take into account the inherent power relations that exist in Israeli society, in which discrimination against Mizrahi-Jews, and de-Arabisation, have been evident.

Perhaps it was getting late, perhaps the alcohol, but as the discussion continued, our friend went even further and claimed that Jewish-Israelis have actually more in common with even anti-Zionist Satmar Hasidic Jews in Brooklyn than with Arab-Palestinian citizens of Israel who live, for example, in Jaffa. This statement left us, again, puzzled: the Satmar Jews do not speak Hebrew; they do not eat 'Israeli food'; they do not dress in an Israeli fashion; they do not even recognise the Israeli state and its symbols; and in fact, when one thinks about it, the Satmars and Jewish-Israelis share hardly anything in common other than imagining a vague belonging to the same 'something', be it people, nation or religion.

We thought about these arguments a number of times in the following days and again upon writing this book and our conclusion, about that discussion and about its dissonances. How come, we pondered, the cultural 'distance' between Tel Aviv to New York, can seem shorter for some than the distance between Tel Aviv and Jaffa? It was a journey, we realised, that needed to be measured not in units of distance, but in perceptions and needs, of genuine and imagined cultures, and of socio-political erasures and constructions.

Jewish-Zionists and Arab-Palestinians: on Influences and Appropriations

As we showed in the book, a number of scholars have raised the subject of Jewish and Arab-Palestinian relations, and in particular the early Zionist settler fascination with Arab and Arab-Palestinian cultures, items and symbols. From Bartal's 'Cossack and Bedouin' to Govrin's 'Enemies or Cousins' analogies, it is clear that early Zionist settlers viewed local Arab-Palestinians, in particular the desert-inhabitant Bedouins and the land-working agricultural fallaḥs, as useful sources to emulate in their desire to become natives in the land of Palestine/*Erets Yisraeli*. They aspired to become *sabras* – to be fruits coming out of the Middle Eastern ground, which are local, not foreign, to the place.

This desire, as we have explained, stemmed from a number of factors. Firstly, the focus on the Bedouins and the fallaḥs, and not the urban Arab-Palestinians for example, was not coincidental. The Bedouins represented for the European-Jewish Zionist immigrants, the strong, noble, uninhibited and free people they wanted to become. The fallaḥs, similarly, exemplified a set of desired values: the hard work, rootedness and devotion to the land that the Zionist immigrants wanted to instil. Alongside these reasons, one should remember that in the process of localisation, the Jewish immigrants needed also to be able to survive in Palestine. Therefore, and from an instrumental point of view, the settlers required skills and knowledge possessed by the Arab-Palestinian locals in order for their endeavour to succeed. On top of this, the Jewish view on the local 'Other' had also spiritual, religious and emotional dimensions. The cultures of the Bedouin and the fallaḥ reminded early settlers of the biblical life and world some of them sought to connect with, and some even went as far as claiming that these two groups shared common ancestry with Jews. Lastly, and connected to the last point, some Jewish immigrants reached the conclusion that if the claim of common ancestry was true, then the adoption of local cultural elements was not based on *appropriation*, but on Jews taking back and rediscovering their own Biblical-historical traditions, and was therefore related to the idea of the Jewish *return* on historical, religious and cultural levels. All in all, we argue that as a result of these factors, early settlers found it useful and suitable to imitate, adopt, adapt and later appropriate local customs, traditions, symbols and words. This was the principal process that we have unearthed in the book, and which changed in style, volume and recognition with time and with the shifting political environment in Palestine/Israel, yet was kept in the DNA of what Jewish-Israelis perceive as 'Israeliness'. It was an ongoing love-hate tango with the Arab-Palestinian 'Other', which on the one hand represented the opposite of the 'self', and on the other hand, its presence was a mandatory ingredient in the creation of many of the customs, traditions and practices considered as local and as *Israeli*.

Reading through most of the literature on the subject, however, one could come to the conclusion that this process of fascination leading on to adaptation and appropriation was important in the early Zionist period and then gradually receded in importance in the run up to 1948. From that point onwards, Jewish-Israelis were no longer fascinated with local Arab-Palestinians – indeed they saw them as the enemy – and no longer required their culture as a resource. In other words, the borrowing of specific Arabic terms and culinary items, by groups such as *Ha-Shomer* and the Palmach, were representative of a particular period which came to an end in 1948. Arab-Palestinians might have influenced early settlers, who chose to dress, eat, talk, and dance in particular ways, but this type of relationship, while important in understanding early and pre-state Zionism and emerging Hebrew culture, was no longer important and does not hold any significant insights into modern Jewish-Israeli identity and culture.

As we have demonstrated, the line of thinking according to which the Arab-Palestinian influence on Hebrew culture has been dramatically reduced following the creation of Israel as an independent state in 1948, is simply inaccurate and does not reflect the reality of Jewish-Arab-Palestinian relations. Not only were the early relations between settlers and Arab-Palestinians important – we would say essential – to our understanding of modern life in Israel and to Jewish-Israeli identity and culture, but the fascination leading to adaptation of Arab and Arab-Palestinian cultures did not end in 1948; it is in fact an ongoing process.

We believe that the study of the early appropriation of Arab-Palestinian culture is still important for understanding modern Jewish-Israeli identity and culture for two main reasons. Firstly, because many of the customs and traditions, which Jewish-Israelis define as belonging to the Israeli way of life and that represent 'Israeliness', are based on those early relations and cultural appropriations. We claim there was perhaps some modification in what is perceived as 'sabra' and 'local' before and after 1948, but not a dramatic shift or change. As a matter of fact, from 1948 onwards, and as we have demonstrated in the previous chapters, there are many case studies indicating the Jewish-Israeli fascination with local Arab-Palestinian people and culture and leading on to imitation and appropriation. This is true for the folk songs and dances Jewish-Israelis are taught at kindergarten and primary school; it is true for many of the food items that epitomise Israeli food culture; it is true for many of the symbols of the state; and it is also true for a considerable number of the words and expressions used in Modern Hebrew.

Secondly, those early, pre-state, relations are important to highlight because of the dynamics and patterns they later instituted and ingrained in Jewish-Israeli society, specifically the relations between Jews and Arab-Palestinians in the state of Israel and between Israel and Arab-Palestinians in the Occupied Palestinian Territories, but also, because they provide an interesting take on Ashkenazi-Mizrahi relations in Israel. While in the early, pre-state, period the adoption and imitation of Arab-Palestinian customs and traditions was not always seen as problematic, this view dramatically changed following 1948. In the aftermath of the 1948 war and the establishment of Israel, the early fascination leading to appropriation had to be rewritten to fit into the official Zionist narrative the state propagated and to the new 'accepted' discursive boundaries of the state. This constructed narrative emphasised the Jewish *return* to their pre-Arab land, the importance of Jewish exclusivity, and the *uniqueness* of the Jewish-Israeli nation, including its culture and history. Any suggestion of an Arab past or an Arab influence was seen as undermining the Jewish-Israeli claim. As a result, Israeli national identity and emerging culture inherently had to be de-Arabised.

This process of rewriting and blurring the past became even more pronounced with the arrival of hundreds of thousands of Jewish immigrants from the Arab world. Their integration into the Israeli state and society was dependent to a certain extent on them denying and forgetting their 'Arab' roots and their adoption of a Jewish-Zionist exclusive identity. In other words, these people, who could have been titled as Arab-Jews before 1948, had had to decide whether they were Arabs *or* Jews following their arrival to Israel, as the two terms were presented as mutually exclusive. The process of de-Arabisation was important therefore also for setting the stage for clear *separation* between Jews and Arabs and the clear presentation and construction of the latter, especially the Arab-Palestinians, as the hostile 'other'. The two people were thus set on separate paths and represented as binaric oppositions that had very little in common.

However, this Zionist view was not based on the reality of Jewish-Arab ongoing relations in Palestine/Israel, before and after 1948, and one of the interesting and important points we wanted to raise and demonstrate in the book is that the process of appropriation (with or without fascination) has been a constant one. The appropriation and cultural diffusion between Jewish-Israelis and Arab-Palestinians has continued after 1948 and up to present day, albeit in different forms and manifestations. From the Israeli gourmet hummus culture to the trendy *jallābiyya* dresses, and from Israeli dances to the Hebrew swear words commonly used in Israel, Jewish-Israeli identity and culture are constantly being shaped and influenced by interactions – real or imagined – with the Arab 'Other' and with the Arab culture.

The continuation of this process raises several challenges for Jewish-Israeli society, in particular how to relate to Arab-Palestinians and their culture and whether this threatens the official Zionist modern narrative that sees the two people as binary opposites. As a result of these issues, and as we have shown, there have been a number of different strategies and approaches adopted by Jewish-Israelis in reference to the Arab-Palestinian influence, or its alleged 'absence'. These have ranged from celebration and recognition of Arab-Palestinian culture to complete denial of cultural influence.

The denial-oriented approach disregards the Arab influence on Israeli culture, an act that fits in and seems 'logical' in light of the Jewish-Zionist discourse. Yet we argue that the same discourse can also explain the other side of the Israeli approach: that of the Israeli new trend of 'celebrating' Arab influence on Israeli culture. As we analyse it, and taking for example Arab-Palestinian food in Israel, the Israeli 'celebration' of local Arab-oriented dishes, ingredients, restaurants and chefs, does not challenge the Israeli discourse, in fact, it might even bolster it. We believe that this celebration is a result of the new power-relations in Israel, and the fact that it is possible and not threatening anymore to acknowledge some of the Arab influences on 'Israeli food'. This is due to the fact that Israeli culture is already inherently 'Israeli', a fact that is very deeply ingrained in many Jewish-Israelis. The celebration of an Arab-Palestinian chef or hummus plate, therefore, is not viewed as threatening. Additionally, this recognition and celebration makes it easier for Jewish-Israelis to ignore the basic, continuous and ongoing Arab influence on Israeli culture generally and on the Israeli food culture more specifically, and the actual relations that include marginalisation and occupation between Jewish-Israelis and Arab-Palestinians. This basic disregard is needed as it also enables Jewish-Israelis to ignore the daily acts of appropriation and colonisation. In other words, it is a celebration that remains well within the Zionist discourse, and does not enable a genuine new outlook on the old Israeli recipe of creating a national Jewish identity using many Arab-Palestinian ingredients.

Israeli Culture and Identity: Palestinian Origins – Acknowledged and Hidden

As our point of departure for this book, we argue that everyday life is grounded, to a large extent, on the importance of habit and routine and the relegation of critical self-reflection to the margins. While framing our lives on perceptions of being part of particular groups and cultures, we tend not to ask where our identities and cultural practices come from or what the norms and values we cherish are based on. Instead, our social lives are often *based* on not asking any of these questions. This is because our ideas regarding identity and culture have been formed through years of institutionalisation, which make us take for granted much of who we are and what we do.

Looking at our case study of Jewish-Israeli national identity and culture, it is evident that Jewish-Israelis tend not to question the prevailing Zionist discourse and the way it relates to issues concerning their identity and culture. Quite the opposite, from early on, Jewish-Israeli narrative required Jewish-Israelis not to question the official Zionist discourse and Israel's 'Jewishness' as these are considered a direct threat to the state and the nation.[1] According to this discourse, the Jewish-Israeli national identity and culture are portrayed and explained as being based on many elements, the majority of them Jewish, but on hardly any Arab ones. The Arab, and specifically the Arab-Palestinian, is presented as a threat, as a source of anxiety, as invading and limiting the Jewish-Israeli space, and as the 'Other' with whom Jewish-Israel have little in common. The truth, however, and as this book shows, is that Jewish-Israeli identity and culture have been shaped in light of ongoing relationships and interactions between Jews in Palestine/Israel and Arab people and culture.

When we discussed in the book Jewish-Israeli identity and culture, we acknowledged that these have had a wide range of influences, among these were also Arab and Arab-Palestinian elements. When we looked at them in greater detail through Israeli food, Israeli dance, Israeli music, or Israeli symbols, we found – somewhere in their very root – also an Arab component. This is a unique influence not only because the Arab-Palestinian influence is common in different cultural fields, but because it seems that these influences are the least noted. For example, we can say there was a strong Yiddish influence on the Hebrew language (take for example Hebrew grammar, and also vocabulary – '*ma nishma*' and '*vos hert zich*' as mere examples)[2] but a much smaller influence of Yiddish-culture food on what is considered 'sabra' and 'Israeli food'. Regarding the latter, this is quite the opposite, as food products like *gefilte fish* or *jellied calfs foot* (lit. '*regel krusha*') are not only not considered 'Israeli' but rather East-European, Jewish and *diasporic* food. In other words, we argue that Arab and Arab-Palestinian influence is much more important in understanding Jewish-Israeli identity and culture than given credit or recognised, and that it had an effect – at times basic and at times more profound – on the *different* cultural fields that constitute what Jewish-Israelis perceive as 'Israeliness' and the Israeli way of life. We believe that due to political reasons, the Arab influence on Israeli culture has been underestimated and overlooked. For many Jewish-Israelis it seems that

---

1   See, for example the rise of groups such as Im Tirzu, which oppose the questioning of Israeli past, history, and try to 'intimidate' post-Zionist scholars and thinkers.

2   See, for example, the following interview with Israeli Linguist, Professor Ghil'ad Zuckerman: 'The revival of the Hebrew Language – Interviews with Ghil'ad Zuckermann and Ugo Volli', November 2013, http://www.informazionecorretta.com/dossier.php?l=en&d=18 (accessed 4 February 2015)

acknowledging this influence threatens to undermine some of the foundations of what is considered the essence of Israeliness and the Israeli way of life.

As we argue, these actions – that of influence and that of disregard – stem from a combination of processes. These include political developments, globalisation-oriented processes, as well as business, media and physical contacts. And all these processes – modern or post-modern, national or post-national – centre around the national need to 'prove' who am I and what is 'mine'. We therefore analyse the Israeli-Palestinian clash during the flower competition in Beijing, the social media uproar over the 'Palestine' gazelle or over products such as Knāfeh and hummus as part and parcel of the same fight or struggle over identity and political location. Furthermore, and as we have shown, identities are fluid concepts, and therefore there is an ongoing struggle over them – even if it includes 'shifts in tactics' to bolster them. These 'tactic changes' included in our case study, the 'return' to Arab food and dress and their meaning – either in uncovering the pre-Arab, Jewish roots, or in attempting to connect Jewish (and Mizrahi) culture to the Arab culture and so to 'blur' the original ownership.[3]

All in all, we argue that the Zionist presentation of the Jewish and Arab identity and culture as two binaries is misleading. The two identities should be viewed more accurately as a scale with overlapping points, while acknowledging that – despite the conflict and at times because of the conflict – it is hard to admit that at the end of many Hebrew sentences sits an Arab smoking a 'nargilah' and that the Arab-Palestinian 'Other' is actually at the very heart of the Jewish-Israeli 'Self'. We believe that it is the *struggle over indigenousness* that needs to be highlighted in any attempt to study and analyse Israeli culture. It is a heated struggle, and a continuous one, which has had a profound influence on the creation of what is perceived to be the 'Israeli culture'. We also think that focusing on this indigenousness-oriented struggle can shed new light on research and studies about Israeli society specifically, but also about societies that are found in the midst of conflict in which there is an asymmetric power relation and fight over land and memory.

Jewish-Israelis and Arab-Palestinians share a number of similarities and points of contact that allow for easier diffusion of culture and symbols. These include, for example the presence of large communities of Jews who have originated in Arab countries and the increasing visibility and involvement of Arab-Palestinians in Israeli politics, economy and society. It is therefore expected that this proximity will result in constant cultural diffusion. However, recognition of this proximity and cultural diffusion threatens the Jewish-Israeli claim of uniqueness and indigenousness. As a result, the Israeli 'acceptance' that 'their' history, identity and culture are not unique and are to an extent a product of their relationship with the Arab-Palestinian people, becomes an uncomfortable

---

3   This also corresponds with the way Benvenisti (2002: 12) analysed the Israelisation of cartography and the creation of the Hebrew Map.

truth for many. Here, we believe, lies a key element in the ongoing conflict in Palestine/Israel, but also a possible breakthrough for its possible solution.

The late Arab-Palestinian poet Mahmoud Darwish once said that in the future, in a time of reconciliation between the Jews and the Arabs, 'the Jew will not be ashamed to find an Arab element in himself, and the Arab will not be ashamed to declare that he incorporates Jewish elements' (quoted in Behar, 2011: 93). Darwish knew well that the struggle of indigenousness, political animosity and conflict between the two people resulted from and in an inability to accept and acknowledge their similarities and the existence of the 'Other' in the 'Self'. As we have shown, with regard to many of the practices and traditions that constitute Israeli identity and culture in the eyes of Jewish-Israelis, one need only to remove the cover to reveal the Palestinian roots that hide – or that were hidden – underneath. This may be intimidating for Jewish-Israelis to admit, but it might be their first genuine attempt to become truly locals.

Bibliography

Abbas, Husam and Nira Rousso (2006). *Lamb, Mint and Pine-Nuts: The Flavours of the Israeli-Arab Cuisine* (Tel Aviv: Yedioth Aharonot) [in Hebrew].

Abufarha, Nasser (2008). 'Land of Symbols: Cactus, Poppies, Orange and Olive Trees in Palestine, Identities' *Global Studies in Culture and Power* 15(3): 343–68.

Abu-Ghosh, Nawal (1996). *The Arab-Israeli Cuisine* (Jerusalem: Keter) [in Hebrew].

Adler, Chaim (2008). 'The Youth Movements and the Construction of the New Society', in Yirmiyahu Yovel (ed.) *New Jewish Time: Jewish Culture in a Secular Age, an Encyclopaedic View* (Jerusalem: Keter) [in Hebrew].

Almog, Oz (2000). *The Sabra: The Creation of the New* Jew (Berkeley, CA: University of California Press).

Almog, Oz (2015). 'Secular Fashion in Israel', in Marzel, Shoshana-Rose and Guy D. Stiebel (eds) *Dress and Ideology: Fashioning Identity from Antiquity to the Present* (London: Bloomsbury).

Almog, Oz and Saltzberg Sima (2010a) 'Hair and Headwear among Women in Religious Zionist Society' *Anashim: People of Israel – Your Guide to Israeli Society* [in Hebrew] 21 January 2010 http://www.peopleil.org/ArticleFiles/7710/7710.pdf (accessed 19 February 2015).

Almog, Oz and Saltzberg Sima (2010b) 'Hair and Headwear patterns among Men in Religious Zionist Society' *Anashim: People of Israel – Your Guide to Israeli Society* [in Hebrew] 21 January 2010 http://www.peopleil.org/ArticleFiles/7708/7708.pdf (accessed 19 February 2015).

Almog, Oz and Saltzberg Sima (2010c) 'Dress Styles among Men in Religious Zionist Society' *Anashim: People of Israel – Your Guide to Israeli Society* [in Hebrew] 21 January 2010 http://www.peopleil.org/ArticleFiles/7647/7647.pdf (accessed 19 February 2015).

Almog, Oz and Saltzberg Sima (2010d) 'Dress Styles among Women in Religious Zionist Society' *Anashim: People of Israel – Your Guide to Israeli Society* [in Hebrew] 21 January 2010 http://www.peopleil.org/ArticleFiles/7646/7646.pdf (accessed 19 February 2015).

Amara, Muhammad (2008). *Teaching Arabic as a Foreign Language in Israeli-Jewish Schools* (Haifa: Jewish-Arab Centre, University of Haifa) [in Hebrew].

Amit-Kochavi, Hannah (2011). 'The people behind the words: Professional profiles and activity patterns of translators of Arabic literature into Hebrew (1896–2009)', in Sela-Seffy, Rakefet and Miriam Shlesinger (eds) *Identity and Status in the Translational Professions* (Amsterdam: John Benjamins BV).

Anderson, Benedict (1983). *Imagined Communities: Reflections on the Origin and Spread of Nationalism* (London: Verso).

Appadurai, Arjun (1988). 'How to Make National Cuisine: Cookbooks in Contemporary India' *Comparative studies in Society and History* 30(1): 3–24.

Arian, Asher (2014). *A Portrait of Israeli Jews: Belief, Observance, and Values of Israeli Jews 2009* (The Israel Democracy Institute and the Avi Chai Israel Foundation) [in Hebrew] available at http://www.idi.org.il/ (accessed 20 December 2014).

Atsuko, Ichijo and Ronald Ranta (2015). *Food, National Identity and Nationalism: From the Everyday to the Global* (Basingstoke: Palgrave Macmillan).

Avieli, Nir (2005). 'Vietnamese New Year Rice Cakes: Iconic Festive Dishes and Contested National Identity' *Ethnology* 44: 167–87.

Avieli, Nir (2013). 'Grilled nationalism: Power, Masculinity and Space in Israeli Barbeques' *Food, Culture and Society* 16(2): 301–20.

Avinery, Yitzhak (1964). *The Occupations of the Hebrew Language in Our Times* (Tel Aviv: Ha-Kibbutz ha-Artzi) [in Hebrew].

Bar-David, Moli (1964). *Folkloric Cookbook: Delights for Israeli Festivals* (Tel Aviv: Bar-David) [in Hebrew].

Bardenstein, Carol (1998). 'Threads of memory and Discourses of Rootedness: of Trees, Oranges and the Prickly-Pear Cactus in Israel/Palestine' *Edebiyat* 8: 1–36.

Bardenstein, Carol (2007). 'Figures of Diasporic Cultural Production: Some Entries from the Palestinian Lexicon' in Marle-Aude Baronian, Stephan Besser and Yolande Jansen (eds) *Diaspora and Memory: Figures of Displacement in Contemporary Literature, Arts and politics* (Amsterdam: Rodopi).

Bar-On, Dan (1999). *The 'Others' Within Us: A Socio-Psychological Perspective on Changes in Israeli Identity* (Ben Gurion University: Israel).

Bar Tal, Daniel and Yona Teichman (2006). *Stereotypes and Prejudice in Conflict: Representations of Arabs in Israeli Jewish Society* (Cambridge: Cambridge University Press).

Bartal, Israel (1976). '"Old Yishuv" and "New Yishuv" – Simile and Reality' *Catedra* 2: 63–76 [in Hebrew].

Barthes, Roland (2008). 'Towards a Psychosiology of Contemporary Food Consumption' in Counihan, Carole and Van Estrik (eds) *Food and Culture: A Reader* (New York: Routledge).

Bat-Yaar, Nurit (2010). *Israel's Fashion Art 1948–2008* (Tel Aviv: Resling) [in Hebrew].

Bell, David and Gill Valentine (1997). *Consuming Geographies: We Are Where We Eat* (New York: Routledge).

Ben-Artzi, Yoiss (2006). 'Alexander Baerwald's Study Tour in Palestine', *Zmanim: A Historical Quarterly*, 96 (Autumn): 14–21.

Ben-David, Orit (2012). 'Tiyul (Hike) as an Act of Consecration of Space' in Eyal Ben-Ari and Yoram Bilu (eds) *Grasping Land: Space and Place in*

Contemporary Israeli Discourse and Experience (Albany: State University of New York Press).

Ben-Ezer, Ehud (1989). 'Landscapes and Borders: A sense of Siege in Israeli Literature' *Shofar* 7(3): 24–31.

Ben-Meir, Orna 2008. 'The Israeli Shoe: Biblical Sandals and Native Israeli Identity', in Edna Nahshon (ed.) *Jews and Shoes* (Oxford: Berg).

Ben-Naeh, Yaron (2005). 'One Cup of Coffee': Ordinance Concerning Luxuries and Recreation: a Chapter in the Cultural and Social History of the Sephardi Community of Jerusalem in the Nineteenth Century' *Turcica* 37: 155–85.

Ben-Rafael, Eliezer and Stephen Sharot (1991). *Ethnicity, Religion, and Class in Israeli Society* (Cambridge: Cambridge University Press).

Benvenisti, Meron (2002). *Sacred Landscape: The Buried History of the Holy Land since 1948* (Berkeley, CA: University of California Press).

Benvenisti, Meron (2012). *The Dream of the White Sabra* (Jerusalem: Keter Books) [in Hebrew].

Ben-Yehouda, Eliezer (1912). 'Filling the missing parts in our language' in Ben-Zion S, Yalin D and Tsifroni A (eds) *Proceedings of the Hebrew Language Council, Vol 4* (Jerusalem: The Hebrew Language Council) [in Hebrew].

Ben-Yehouda, Eliezer (1980). *A Complete Dictionary of Ancient and Modern Hebrew* (Jerusalem: Makor) [in Hebrew].

Ben-Yehouda, Netiva (1981). *1948 – Between the Calendars* [in Hebrew] (Jerusalem: Keter).

Ben-Zvi, Yitzhak (1975). *Erets Yisrael and its Settlement during the Ottoman Era* (Jerusalem: Yad Yitzhak Ben Zvi).

Behar, Almog (2011)., 'Mahmoud Darwish: Poetry's state of siege'. *Journal of Levantine Studies* 1: 189–99.

Bhabha, Homi K. (1997). *The Location of Culture* (London: Routledge).

Billig, Levy and Avinoam Yellin (1931). *The Arabic Reader: Edited with Notes and a Glossary* (Jerusalem and London: Macmillan).

Billig, Michael (1995). *Banal Nationalism* (London: Sage).

Blau, Joshua (1976). *On the Revival of Modern Hebrew and of Literary Arabic* (Jerusalem: Hebrew Language Academy) [in Hebrew].

Blau, Joshua (1981). *The Emergence and Linguistic Background of Judaeo-Arabic: A Study of the Origins of Middle Arabic* (Jerusalem: Ben-Zvi Institute).

Blommaert, Jan (ed.) (1999). *Language Ideological Debates* (New York: Mouton de Gruyter).

Bluestein, Gene (1989). *Anglish/Yinglish: Yiddish in American Life and Literature* (Athens, GA: The University of Georgia Press).

Bourdieu, Pierre (1992). *Language and Symbolic Power* (Cambridge: Polity Press).

Boullata, Kamal (2001). 'Asim Abu Shaqra: The artist eye and the cactus tree' *Journal of Palestine Studies* 30(4): 68–82.

Boyd, Lee and Victor Kaftal (2000). *Canaan Dog: A Complete and Reliable Handbook* (Neptune City, NJ: TFH Publications).

Byman, Daniel (2011). *A High Price: The Triumphs and Failures of Israeli Counterterrorism* (Oxford: Oxford University Press).

Chaver, Yael (2004). *What Must be Forgotten: The Survival of Yiddish Writing in Zionist Palestine* (Syracuse: Syracuse University Press).

Cohen, Adir (1985). *The Ugly Face in the Mirror: the Arab-israeli Conflict in Hebrew Children's Books* (Tel Aviv: Reshafim) [in Hebrew].

Cohen, Hillel (2013). *1929: Year Zero of the Israel-Palestine Conflict* (Tel Aviv: Keter) [in Hebrew].

Cohen, Mark, R. (1994). *Under Crescent and Cross: The Jews in the Middle Ages* (Princeton: Princeton University Press).

Confino, Alon (1997). *The Nation as a Local Metaphor: Wurttemberg, Imperial Germany, and National Memory 1871–1918* (Chapel Hill: University of North Carolina Press).

Cornfeld, Lilian (1949). *What to Cook with the Austerity Portions* (Tel Aviv: Dov Gutman Press) [in Hebrew].

Dagmar, G. and W. Reuschel (1992). 'Status Types and Status Changes in the Arabic Language' in U. Ammon and M. Hellinger (eds) *Status Change of Languages* (Berlin: de Gruyter).

Dahl, Gudrun (1998). 'Wildflowers, Nationalism and the Swedish Law of Commons' *Worldviews: Environment, Culture, Religion and Ecology* 2(3): 281–302.

Dolev, Dana and Haim Gordan (1992). 'Architectural Orientalism in early Zionist Buildings: the Case of the Hebrew University', *The Centennial Review*, 36(2): 361–72.

Don, Ellis and Ifat Maoz (2002). 'Cross-cultural Argument Interactions Between Israeli-Jews and Palestinians' *Journal of Applied Communications Research* 30: 181–94.

Don-Yehiya, Eliezer and Charles S. Liebman (1981). 'The Symbol System of Zionist-Socialism: An Aspect of Israeli Civil Religion' *Modern Judaism* 1(2): 121–48.

Dowty, Alan (2008). *Israel Palestine* (Cambridge: Polity).

Dror, Yuval (2008). 'Pioneers of Education', *The Israeli Association of Teachers Journal* (December 2008): http://www.itu.org.il/?CategoryID=1476&Article ID=12507 (accessed: 21 April 2015) [in Hebrew].

Earle, Rebecca (2007). *The Return of the Native: Indians and Myth making in Spanish America, 1810–1930* (Durham, NC: Duke University Press).

Edensor, Tim (2002). *National Identity, Popular Culture and Everyday Life* (Oxford: Berg).

Elgenius, Gabriella (2011). *Symbols of Nations and Nationalism: Celebrating Nationhood* (Basingstoke: Palgrave Macmillan).

Eisenstadt, S.N. (1951). 'Youth, Culture and Social Structure in Israel' *The British Journal of Sociology* 2(2): 105–14.

Eisenstadt, Shmuel Noah (2002). *Jewish Civilization: The Jewish Historical Experience in a Comparative Perspective and its Manifestations in Israeli*

Society (Beersheba: The Ben-Gurion Heritage Institute – Ben-Gurion University of the Negev) [in Hebrew].

Elboim-Dror, Rachel (1990). *The Hebrew Education in Erets Israel* (Jerusalem: Ben-Zvi Institute) [in Hebrew].

Eliram, Talila (2006). *Come, Thou Hebrew Songs: the Songs of the Land of Israel – Musical and Social Aspects* (Haifa: Haifa University Press) [in Hebrew].

El Or, Tamar (2012). 'The Soul of the Biblical Sandal: On Anthropology and Style' *American Anthropologist* 114(3): 433–45.

El Or, Tamar (2014). *Sandals: The Anthropology of Local Style* (Tel Aviv: Am Oved) [in Hebrew].

Even-Zohar, Itamar (1981). 'The Emergence of a Native Hebrew Culture in Palestine: 1882–1948' *Studies in Zionism: Politics, Society, Culture* 2(2): 167–84.

Eyal, Gil (2006). *The Disenchantment of the Orient: Expertise in Arab Affairs and the Israeli State* (Stanford, CA: Stanford University Press).

Forbes, Andrew and David Henley (2012). *People of Palestine* (Chiang Mai: Cognoscenti Books).

Fanon, Franz (1991). *Black Skin, White Masks* (London: Pluto Press).

Fox, Jon E. and Miller-Idriss, Cynthia (2008). 'Everyday Nationhood' *Ethnicities*, 8(4): 536–76.

Frank, Daniel (ed.) (1995). *The Jews of Medieval Islam: Community, Society, and Identity* (Leiden: E.J. Brill).

Freudenthal, Gad (2011). 'Arabic and Latin Cultures as Resources for the Hebrew Translation Movement: Comparative Considerations, Both Quantitative and Qualitative' in Gad Freudenthal (ed.) *Science in Medieval Jewish Cultures* (New York: Cambridge University Press).

Friedman Menachem (1988). *Society and Religion: The non-Zionist Orthodoxy in the Land of Israel* (Jerusalem: Ben Zvi Institute) [in Hebrew].

Galai, Binyamin (1988). *The Words in Our Life* (Tel Avi: Dvir/Zmora Bitan) [in Hebrew].

Gardner, R.C. and W.E. Lambert (1972). *Attitudes and Motivations in Second Language Learning* (Rowley: Newbury House).

Geislar, Michale E. (2005). 'Introduction: What are National Symbols and What do They Do to US?' in Geislar, Michael E. (ed.) *National Symbols, Fractured Identities: Contesting the National Narrative* (Lebanon: University Press of New England).

Gellner, Ernst (1997). *Nationalism* (New York: New York University Press).

Gera, Gershon (1985). *The Ha-Shomer Book* (Tel Aviv: Israeli Ministry of Defence) [in Hebrew].

Gilroy, Paul (2002). *There Ain't No Black in the Union Jack: The Cultural Politics of race and Nation* (London: Routledge).

Gitelman, Zvi (1977). 'Absorption of Soviet Immigrants' in Curtis, Michael and Chertoff, Mordecai S. (eds) *Israel: Social Structure and Change* (New Jersey: Transaction Books).

Goitein, S.D. (1955). *Jews and Arabs: Their Contacts through the Ages* (New York: Schoken Books).

Golden, Deborah (2005). 'Nourishing the Nation: The Uses of Food in an Israeli Kindergarten' *Food and Foodways: Explorations in the History and Culture of Human Nourishment* 13(3): 181–99.

Gorni, Yosef (1987). *Zionism and the Arabs, 1882–1948* (Oxford: Oxford University Press).

Govrin, Nurit (1989). 'Enemies or Cousins? … Somewhere in Between, The Arab Problem and its Reflection in Hebrew Literature: Developments, Trends, and Examples' *Shofar* 7(3): 13–23.

Grant, Elihu (1907). *The Peasantry of Palestine* (New York: Pilgrim Press).

Gribetz, Jonathan (2013). 'Mustaʿribūn', *Encyclopedia of Jews in the Islamic World* Executive Editor: Norman A. Stillman (Brill online).

Griffith, Sidney, H. (2013). *The Bible in Arabic: The Scriptures of the 'People of the Book' in the Language of Islam* (Princeton: Princeton University Press).

Gur, Janna (2008). *Fresh Flavours from Israel* (Tel Aviv: Al Hashulchan Gastronomic Media).

Gvion, Liora (2012). *Beyond Hummus and Falafel: Social and Political Aspects of Palestinian Food in Israel* (London: University of California Press).

Hall, Stuart and Paul du Gay (eds) (1996). *Questions of Cultural Identity* (London: Sage).

Halperin, Liora (2006). 'Orienting Language: Reflections on the Study of Arabic in the Yishuv' *Jewish Quarterly Review* 96(4): 481–9.

Halperin, S. (1970). *Dr. A. Biram and his 'Reali' School: Tradition and Experimentation in Education* (Jerusalem: R. Mass) [in Hebrew].

Handelman, Don (2004). *Nationalism and the Israeli State: Bureaucratic Logic in Public Events* (Oxford: Berg).

Harshav, Benjamin (1990). *The Meaning of Yiddish* (Berkeley, CA: University of California Press).

Hearn, Jonathan (2006). *Rethinking Nationalism: A Critical Introduction* (Basingstoke: Palgrave Macmillan).

Heldke, Lisa (2003). *Exotic Appetites: Ruminations of a Food Adventurer* (London: Routledge).

Herzl, Theodor (1972). *The Jewish State: An Attempt at a Modern Solution of the Jewish Question* [1896] (London: H. Pordes).

Herzl, Theodor (1997). *Altneuland – The Old-New Land* [1902] (Tel Aviv: Babel Publishers) [in Hebrew].

Hever, Hannan (1994). 'Territoriality and Otherness in Hebrew Literature of the War of Independence' in Laurence Jay Silberstein and Robert L. Cohn (eds) *The Other in Jewish Thought and History: Constructions of Jewish Culture and Identity* (New York: New York University Press).

Hinnawi, Miriam (2006). *Arab Cuisine from the Galilee* (Jerusalem: Academon) [in Hebrew].

Hirsch, Dafna (2011). 'Hummus is best when it is fresh and made by Arabs: The gourmetization of hummus in Israel and the return of the repressed Arab' *American Ethnologist* 38 (4): 617–30.

Hirsch, Dafna (2014). 'Hygiene, Dirt and the Shaping of a New Man among the Early Zionist Halutzim' *European Journal of Cultural Studies* (published online 15.4).

Hirsch, Dafna and Ofra Tene (2013). 'Hummus: The Making of an Israeli Culinary Cult' *Journal of Consumer Culture* 13 (1): 25–45.

Hirshberg, Jehoash (1995). *Music in the Jewish Community of Palestine 1880–1948. A Social History* (Oxford: Clarendon Press).

Hixson, Walter L. (2013). *American Settler Colonialism: A History* (New York: Palgrave Macmillan).

Hobsbawm, E.J. (1983). *Nations and Nationalism Since 1780* (Cambridge: Cambridge University Press).

Hornberger, Nancy J. (2003). 'Multilingual Language Policies and the Continua of Biliteracy: An Ecological Approach' in N.J. Hornberger (ed.) *Continua of Biliteracy: An Ecological Framework for Educational Policy, Research and Practice in Multilingual Settings* (Clevedon: Multilingual Matters).

Hutchinson, John (2004). 'Myth Against Myth: The Nation as Ethnic Overly' *Nations and Nationalism* 10(1/2): 109–23.

Johnson, Paul (1987). *History of the Jews* (New York: Harper and Row).

Kadman, Gurit (1952). 'Yemenite Dances and Their Influence on the New Israeli Folk Dances' *Journal of the International Folk Music Council* 4: 27–30.

Kadman, Noga (2008). *On the Side of the Road in the Margins of Consciousness: The Depopulated Palestinian Villages of 1948 in the Israeli Discourse* (Jerusalem: November Books) [in Hebrew].

Karlinsky, N. (2000). *Citrus Blossoms: Jewish Entrepreneurship in Palestine, 1890–1939* (Jerusalem: Magnes Press) [in Hebrew].

Karlinsky, N. (2005). *California Dreaming: Ideology, Society and Technology in the Citrus Industry of Palestine, 1890–1939* (Albany: State University of New York Press).

Kark, R. (1990). *Jaffa: A City in Evolution 1799–1917* (Jerusalem: Yad Izhak Ben-Zvi Press).

Katriel, Tamar (1986). *Talking Straight: Dugri Speech in Israeli Sabra Culture* (Cambridge: Cambridge University Press).

Katriel, Tamar (1987). 'Rhetoric in Flames: Fire Inscriptions in Israeli Youth Movement Ceremonies' *Quarterly Journal of Speech* 73(4): 444–59.

Katzburgm Nathaniel (1996). 'The Educational Controversy in the Old Yishuv', in *Heichal Shlomo Yearbook* (Jerusalem: Heichal Shlomo) [in Hebrew].

Khalidi, Rashid (2010). *Palestinian Identity: The Construction of Modern National Consciousness* (New York: Columbia University Press).

Khazzoom, Aziza (2003). 'The Great Chain of Orientalism: Jewish Identity, Stigma Management, and Ethnic Exclusion in Israel' *American Sociological Review* 68(4): 481–510.

Kimmerling, Baruch (2004). *Immigrants, Settlers, Natives: The Israeli State and Society between Cultural Pluralism and Cultural Wars* (Tel Aviv: Am Oved) [in Hebrew].

Kimmerling, Baruch and Joel S. Migdal (2009). *The Palestinian People: A History* (Cambridge, MA: Harvard University Press).

Kinberg, Naphtali and Rafael Talmon (1994). 'Learning of Arabic by Jews and the use of Hebrew among Arabs in Israel' *Indian Journal of Applied Linguistics* 20(1–2): 37–54.

Kitto, John (1841). *Palestine: The Physical Geography and Natural History of the Holy Land* (London: Charles Knight and Co.).

Koenig, Samuel (1952). 'Israeli Culture and Society' *American Journal of Sociology* 58(2): 160–66.

Kolsto, Pal (2006). 'National Symbols as Signs of Unity and Division' *Ethnic and Racial Studies* 29(4): 676–701.

Korn, Yitzhak (1983). *Jews at the Crossroad* (East Brunswick, NJ: Cornwall Books).

Kosover, M. (1966). *Arabic Elements in Palestinian Yiddish: The Old Ashkenazic Jewish Community in Palestine, its History and its Language* (Jerusalem: R. Mass).

Kraemer, Roberta (1990). 'Social psychological Factors related to the Study of Arabic among Israeli Jewish High School Students' (Unpublished PhD. Thesis, School of Education, Tel Aviv University).

Kroyanker, David (1996). *Jerusalem: Neighbourhoods and Houses, Periods and Style* (Jerusalem: Keter) [in Hebrew].

Kroyanker, David (2006). 'More Oriental than the Orient: the Language of Pastiche in Jerusalem', *Zmanim: A Historical Quarterly*, 96 (Autumn): 28–37 [in Hebrew].

Laor, Dan (2003). *Nathan Alterman: A Biography* (Tel Aviv: Am Oved) [in Hebrew].

Laqueur, Walter (1972). *A History of Zionism* (New York: Schoken Books).

Levi-Strauss, Claude (1986). *The Raw and The Cooked* (London: Penguin).

Levine, Mark (2005). *Overthrowing Geography: Jaffa, Tel Aviv, and the Struggle for Palestine* (Berkeley, CA: University of California Press).

Levin, Michael (1980). 'Outline of the Trends of the Emergence and Crystallization of Local Art and Architecture in Eretz-Israel', *Cathdera: For the History of Eretz Israel and its Yishuv*, 16: 194–204 [in Hebrew].

Levin, Michael (2006). 'Five Attitudes to the Orient in Israeli Architecture', *Zmanim: A Historical Quarterly*, 96 (Autumn): 38–47.

Levy, Lital (2014). *Poetic Trespass: Writing between Hebrew and Arabic in Israel/Palestine* (Princeton: Princeton University Press).

Lewis, Bernard (1987). *The Jews of Islam* (Princeton: Princeton University Press).

Litani, Yehuda and Naim Araidi (2000). *Not by Hummus Alone: Hummus, Olive Oil, References* (Tel Aviv: Dinur and Modan) [in Hebrew].

Lockman, Zachary (1993). 'Railway Workers and Relational History: Arabs and Jews in British Ruled Palestine' *Society for Comparative Study of Society and History* 35(3): 601–27.

Lockman, Zachary (1996). *Comrades and Enemies: Arab and Jewish Workers in Palestine 1906–1948* (Berkeley, CA: University of California Press).

Maalouf, Tony (2003). *Arabs in the Shadow of Israel: The Unfolding of God's Prophetic Plan for Ishmael's Line* (Grand Rapids, MI: Kregel Publications).

Makoni, Sinfree (1998). 'African languages as European scripts: The shaping of communal memory' in S. Nuttall and and C. Cotzee (eds) *Negotiating the Past: The Making of Memory in South Africa* (Capetown: Oxford University Press).

Manor, Dalia (2002). 'The Dancing Jew and Other Characters: Art in the Jewish Settlement in Palestine During the 1920s' *Journal of Modern Jewish Studies*, 1(1): 73–89.

Manor, Dalia (2005). *Art in Zion: The Genesis of Modern National Art in Jewish Palestine* (London: Routledge).

Mar'i, Abd el-Rahman (2013). *Walla Bseder: A Linguistic Profile of the Israeli Arabs* (Jerusalem: Keter) [in Hebrew].

Marks, Gil (2010). *Encyclopaedia of Jewish Food* (New Jersey: John Wiley and Sons).

Masterman, E.W.G. (1901). 'Food and its Preparation in Modern Palestine' *The Biblical World* 17(6): 407–19.

Mayar, Tamar (2005). 'National Symbols in Jewish Israel: Representation and Collective Memory' in Michael E. Geislar (ed.) *National Symbols, Fractured Identities: Contesting the National Narrative* (Lebanon: University Press of New England).

Mehta-Jones, Shilpa (2005). *Life in Early Ancient Mesopotamia* (New York: Crabtree Publishing Company).

Menashe, Geffen (1991). *Transformation of Motifs in Folklore and Literature* (Jerusalem: R. Mass) [in Hebrew].

Mendel, Yonatan (2013a). 'Re-Arabising the De-Arabised: The Mista'aravim Unit of the Palmach', in A. Bernard, Z. Elmarsafy, and D. Attwell (eds) *Orientalism: Thirty Years Later* (London: Palgrave Macmillan).

Mendel, Yonatan (2013b) 'A Sentiment-Free Arabic: On the Creation of the Israeli Accelerated Arabic Language Studies Programme' *Middle Eastern Studies* 49(3): 383–401.

Mendel, Yonatan (2013c) 'New Jerusalem' *New Left Review* 81: 34–56.

Mendel, Yonatan (2014). *The Creation of Israeli Arabic: Security and Politics in Arabic Studies in Israel*, Palgrave Studies in Languages at War (London: Palgrave Macmillan).

Mendel, Yonatan (2015a) 'Arabic Language in Israel', *Mafte'akh: Lexical Review of Political Thought* 9 (Spring) [in Hebrew]: http://mafteakh.tau.ac.il/2015/04/%D7%94%D7%A9%D7%A4%D7%94-%D7%94%D7%A2%D7%A8%D7%91%D7%99%D7%AA/ (accessed: 23 April 2015).

Mendel, Yonatan (2015b, forthcoming) 'From German philology to local usability: The emergence of "practical" Arabic in the Hebrew Reali School in Haifa', *Middle Eastern Studies*.

Mendel Yonatan and Ronald Ranta (2014). 'Consuming Palestine: Palestine and Palestinians in Israeli Food Culture' *Ethnicities* 14(3): 412–35.

Mey, Zaki (2008). *The Legacy of Tutankhamun: Art and History* (Cairo: Farid Atiya Press).

Meyer, Erna (1937). *How to Cook in Palestine* (Tel Aviv, WIZO) [in Hebrew, English and German].

Mikha Rinat, Mikha Livneh and Oz Almog (2009). 'Development of dress patterns and external appearance in the kibbutz', *Anashim: People of Israel – Your Guide to Israeli Society* [in Hebrew]: http://www.peopleil.org/details. aspx?itemID=7839 (accessed 23 April 2015).

Milson, Menahem (1996). 'The beginnings of Arabic and Islamic Studies at the Hebrew University of Jerusalem' *Judaism* 45(2): 169–83.

Mintz, Sidney W. (1985). *Sweetness and Power: The Place of Sugar in Modern History* (New York: Penguin Books).

Mohammed, Zakaria (2006). 'The Thorn and the Flower' in Philipp Misselwitz and Tim Rientiets (eds) *City of Collision: Jerusalem and the Principles of Conflict Urbanism* (Basel: Birkhauser).

Montanari, Massimo (2004). *Food is Culture* (New York: Columbia University Press).

Monterescu, Daniel (2009). 'To Buy or Not to Be: Trespassing the gated Community' *Public Culture* 21(2): 403–30.

Morahg, Gilead (1986). 'New Images of Arabs in Israeli Fiction' *Prooftexts* 6(2): 147–62.

Moreh, Shmuel (2001). 'The Study of Arabic Literature in Israel' Ariel – The Israel Review of Arts and Letters: Cultural and Scientific Relations Division – Ministry of Foreign Affairs (December).

Morris, Benny (2004). *The Birth of the Palestinian Refugee Problem Revisited* (Cambridge: Cambridge University Press).

Narayan U. (1995). 'Eating cultures: Incorporation, identity and Indian food' *Social Identities* 1(1): 63–86.

Natour, Salman (2009). 'Safar ʿala Safar' in Salman Natour (trilogy in Arabic): *Sixty Years: Walking in the Desert* (Ramalla: Dār al-Shurūq) [in Arabic].

Nocke, Alexandra (2006). *The Place of the Mediterranean in Modern Israeli Identity* (Leiden: Brill).

Or, Iair (2015 forthcoming). *Creating a Style for a Generation': Language Beliefs in the Discussions of the Hebrew Language Committee in the years 1912–1928* [in Hebrew].

Outhwaite, Ben (forthcoming). 'Lines of communication: Medieval Hebrew letters of the 11th century' in E-M. Wagner, B. Outhwaite and B. Beinhoff (eds) *Scribes as Agents of Language Change* (Berlin: De Gruyter).

Oppenheimer, Yochai (1999). 'The Arab in the Mirror: The Image of the Arab in Israeli Fiction' *Prooftexts* 19: 205–34.

Oz, Amos (2002). *A Tale of Love and Darkness* (Jerusalem: Keter) [in Hebrew].

Ozkirmili, Umut (2005). *Contemporary Debates on nationalism: A Critical Engagement* (Basingstoke: Palgrave Macmillan).

Palika, Liz (2007). *The Howell Book of Dogs: The Definitive Reference to 300 Breeds and Varieties* (Hoboken: Wiley Publishing).

Palmer, Catherine (1998). 'From Theory to Practice: Experiencing the Nation in Everyday Life' *Journal of Material Culture* 3: 175–99.

Pappé, Ilan (2006). *The Ethnic Cleansing of Palestine* (Oxford: Oneworld).

Pavlenko, Aneta (2003). '"Language of the Enemy": Foreign Language Education and National Identity' *International Journal of Bilingual Education and Bilingualism* 6(5): 313–31.

Pedhazur, Ami (2012). *The Triumph of Israel's Radical Right* (New York: Oxford University Press).

Peled-Elhanan, Nurit (2012). *Palestine in Israeli School Books: Ideology and Propaganda in Education* (London: I.B. Tauris).

Peleg, Muli (1997). *Spreading the Wrath of God: From Gush Emunim to Rabin Square* (Tel Aviv: Ha-Kibbutz Ha-Meuchad) [in Hebrew].

Pennycook, Alastair (2004). 'Performativity and Language Studies', *Critical Inquiry in Language Studies: An International Journal* 1(1): 1–19.

Penslar, Derek (2007). *Israel in History: The Jewish State in Comparative Perspective* (Abingdon: Routledge).

Peretz-Rubin, Pascal (1987). *Israel Flavours* (Ramat Gan: Ruth Sirkis Publishers) [in Hebrew].

Rabinowitz, Dan and Daniel Monterescu (2008). 'Reconfiguring the "mixed town": urban transformations of ethnonational relations in Palestine and Israel' *International Journal of Middle Eastern Studies* 40(2): 195–226.

Ranta, Ronald and Atsuko Ichijo (2015). *Food, National Identity and Nationalism: From Everyday to Global Politics* (Basingstoke: Palgrave Macmillan).

Ranta, Ronald (2015, forthcoming) 'Re-Arabising Israel's Food Culture' *Food, Culture and Society*.

Raviv, Yael (2002). *Recipe for a Nation: Cuisine, Jewish Nationalism, and the Israeli State.* (Unpublished PhD. dissertation, New York University).

Raviv, Yael (2003). 'Falafel: A National Icon' *Gastronomica: the Journal of Food and Culture* 3(3): 20–25.

Raz, Ayala (1996). *Changing Styles: Hundred Years of Fashion in Israel* (Tel Aviv: Yediot Aharonoth).

Raz-Krakotzkin, Amnon (1993). 'Exile Within Sovereignty: Towards a Critique of the 'Negeation of Exile' in Israeli Culture' *Theory and Criticism* 4: 23–56 [in Hebrew].

Raz-Krakotzkin, Amnon (2005). 'The Zionist Return to the West and the Mizrahi Jewish Perspective' in Kalmar, Iavan Davidson and Derek J. Penslar (eds) *Orientalism and the Jews* (Lebanon: Brandeis University Press).

Regev, Motti (1995). 'Present Absentee: Arab Music in Israeli Culture' *Public Culture* 7: 433–45.

Regev Motti (2000). 'To Have a Culture of our own: on Israeliness and its variants' *Ethnic and Racial Studies* 23(2): 223–47.

Regev, Motti and Edwin Seroussi (2004). *Popular Music and National Culture in Israel* (Berkeley, CA: University of California Press).

Roden, Claudia (1999). *The Book of Jewish Food: An Odyssey from Samarkand and Vilna to the Present Day* (London: Penguin).

Rogers, Eliza Mary (1865). *Domestic Life in Palestine* (Cincinnati: Poe & Hitchcock).

Rogers, Ben (2003). *Beef and Liberty: Roast Beef, John Bull and the English Nation* (London: Vintage).

Roginsky, Dina (2006). 'Nationalism and Ambivalence: Ethnicity, Gender and Folklore as Categories of Otherness' *Patterns of Prejudice* 40(3): 237–58.

Rognisky, Dina (2007). 'Folklore, Folklorism, and Synchronization: Preserved-Created Folklore in Israel' *Journal of Folklore Research* 44(1): 41–66.

Rozen, Minna (1980). 'The Position of the Musta'rabs in the Inter-community Relationships in Erets Israel from the End of the 15th Century to the End of the 17th Century' *Cathedra* 17: 73–101 [in Hebrew].

Rozin, Orit (2006). 'Food, Identity, and Nation-Building in Israel's Formative Years' *Israel Studies Forum* 21(1): 52–80.

Sachar, Howard Morley (1961). *Aliya: The People of Israel* (Cleveland: World Publishing Company).

Said, Edward (1995). *Orientalism: Western Conceptions of the Orient* (London: Penguin Books).

Scholch, Alexander (1981). 'The Economic Development of Palestine, 1856–1882' *Journal of Palestine Studies* 10(3): 35–58.

Segev, Tom (1999). *One Palestine, Complete: Jews and Arabs under the British Mandate* (New York: Metropolitan Books).

Sela-Sheffy, Rakefet (2004). 'What makes one an Israeli? Negotiating identities in everyday representations of "Israeliness"' *Nations and Nationalism* 10(4): 479–97.

Shafir, Gershon (1996). *Land, Labour and the Origins of the Israeli-Palestinian Conflict, 1882–1914* (London: University of California Press).

Shalev, Meir (1988). *Russian Novel* (Tel Aviv: Am Oved) [in Hebrew].

Shapira, Anita (1992). *Land and Power: The Zionist Resort to Force, 1881–1948* (Stanford, CA: Stanford University Press).

Shenhav, Yehouda (2006). *The Arab Jews: A Postcolonial Reading of Nationalism, Religion, and Ethnicity* (Stanford, CA: Stanford University Press).

Shenhav, Yehouda (2012). 'The Politics and Theology of Translation: How do we translate Nakba from Arabic to Hebrew' *Israeli Sociology* 14: 157–84 [in Hebrew].

Shirkis, Othman (2008). 'The Flower of Cyclamen is Endangered', *Environment and Development* (2): http://www.maan-ctr.org/magazine/Archive/Issue2/torath/torath1.htm (accessed: 25 October 2015) [in Arabic].

Shlaim, Avi (2001). *The Iron Wall: Israel and the Arab World* (New York: W.W. Norton).

Shlush, Aharon (1991). *From Jellabiya to Tembel Hat: a Story of a Family* (Tel Aviv: Bnei Shaul) [in Hebrew].

Shohamy, Elana (2006). *Language Policy: Hidden Agendas and New Approaches* (New York: Routledge).

Shohat, Ella (1988). 'Sephardim in Israel: Zionism from the Standpoint of Its Jewish Victims' *Social Text* 19/20: 1–35.

Shohat, Ella (2005). *Israeli Cinema: East/West and the Politics of Representation* (Raanana: The Open University of Israel) [in Hebrew].

Silverman, Eric (2013). *A Cultural History of Jewish Dress* (London: Bloomsbury Academic).

Sirkis, Ruth (1975). *Popular Food from* Israel (Tel Aviv: Zmora Bitan Modan-Publishers).

Skey, Michael (2011). *National Belonging and Everyday Life: The Significance of Nationhood in an Uncertain World* (Basingstoke: Palgrave Macmillan).

Smith, Anthony (1993). *A National Identity* (Reno: University of Nevada Press).

Smith, Anthony (2002). 'When is a Nation' *Geopolitics*, 7 (2): 5–32.

Smith, Anthony (2009). *Ethno-Symbolism and Nationalism: A Cultural Approach* (Abingdon: Routledge).

Snir, Reuven (2005). *Arabness, Jewishness, Zionism: A Struggle of Identities in the Literature of Iraqi Jews* (Jerusalem: Ben-Zvi Institute) [in Hebrew].

Snir, Reuven (2006). '"Ana min al-Yahud": The Demise of Arab-Jewish Culture in the Twentieth Century' *Archiv Orientální* 74: 387–424.

Sorek, Tamir (2004). 'The Orange and the 'Cross in the Crescent': Imagining Palestine in 1929' *Nations and Nationalism* 10(3): 269–91.

Soshuk, Levi and Azriel Eisenberg (eds) (1984). *Momentous Century: Personal and Eyewitness Accounts of the Rise of the Jewish Homeland and State 1875–1978* (Cranbury, NJ: Cornwall Books).

Spolsky, Bernard (1999). 'Language in Israel: Policy, Practice and Ideology', *Georgetown University Round Table on Language and Linguistics*.

Spolsky, Bernard and Elana Shohamy (1999). *The Languages of Israel: Policy, Ideology, and Practice* (Clevedon: Multilingual Matters).

Steinschneider, Moritz (2008). *Jewish Arabic Literature* (Piscataway, NJ: Gorgias Press).

Suleiman, Yasir (1994). 'Nationalism and the Arabic Language: A Historical Overview', in Yasir Suleiman (ed.) *Arabic Sociolinguistics: Issues and Perspectives* (Richmond: Curzon Press, 1994), pp. 3–24.

Suleiman, Yasir (2003). *The Arabic Language and National Identity: A Study in Ideology* (Edinburgh: Edinburgh University Press).

Suleiman, Yasir (2004). *A War of Words: Language and Conflict in the Middle East* (Cambridge: Cambridge University Press).

Suleiman, Yasir (2011). *Arabic, Self and Identity: A Study in Conflict and Displacement* (Oxford: Oxford University Press).

Suleiman, Yasir (2013). *Arabic in the Fray: Language Ideology and Cultural Politics* (Edinburgh: Edinburgh University Press).

Tene, Ofra (2002). *Thus You Shall Cook! Readings in Israeli Cookbooks* (MA thesis, Tel Aviv University) [in Hebrew].

Tessler, Mark (1994). *A History of the Israeli-Palestinian Conflict* (Bloomington: Indiana University Press).

Tessler, Shmulik (2007). *Songs in Uniform: The Story of the IDF Ensembles* (Jerusalem: Yad Ben Zvi) [in Hebrew].

Tilly, Charles (1985). 'War-Making and State-Making as Organised Crime' in Peter B. Evans, Dietrich Reuschmeyer and Theda Skocpol (eds) *Bringing the State Back In* (Cambridge: Cambridge University Press).

Torstrick, Rebecca, L. (2004). *Culture and Customs of Israel* (Westport, CT: Greenwood Press).

Trevor-Roper, Hugh (1998). 'The Invention of Tradition: The Highland Tradition of Scotland' in Eric Hobsbawm and Terence Ranger (eds) *The Invention of Tradition* (Cambridge: Cambridge University Press).

Tzaban, Yair (2008). 'The Youth Movements in Israel: Historical Survey', in Yirmiyahu Yovel (ed.) *New Jewish Time: Jewish Culture in a Secular Age, an Encyclopaedic View* (Jerusalem: Keter) [in Hebrew].

Urian, Dan (1992). 'The Image of the Arab in Israeli Theatre – from Competition to Exploitation (1912–1990)' *Theatre Research International* 17: 46–54.

Urian, Dan (1997). *The Arab in Israeli Drama and Theatre* (Amsterdam: Overseas Publishers Association).

Urian, Dan (2005). 'The Emergence of the Arab Image in Israeli Theatre, 1948–1982' *Israeli Affairs* 1(4): 101–27.

Uzelac, Gordana (2002). 'When is the Nation? Constituent Elements and Processes' *Geopolitics* 7(2): 33–52.

Veracini, Lorenzo (2010). *Settler Colonialism: a Theoretical Overview* (Basingstoke: Palgrave Macmillan).

Weitz, Yechiam (ed.) (2000). *Palmach: Two Sheaves and a Sword* (Ramat Efal: Ministry of Defense Press) [in Hebrew].

Weizman, Eyal (2007). *Hollow Land: Israel's Architecture of Occupation* (London: Verso).

Williams, Raymond (1983). *Culture and Society 1780–1950* (New York: Columbia University Press).

Wilk, Richard (2006). *Home Cooking in the Global Village: Caribbean Food From Buccaneers to Ecotourists* (New York: Berg).

Weinrich, Max (2008). *History of the Yiddish Language* (New Haven, CT: Yale University Press).

Weitz, Yechiam (ed.) (2000). *The Palmach: Two Sheaves and a Sword* (Ramat Efal: Ministry of Defense Press) [in Hebrew].

Yacobi, Haim (2008). 'Architecture, Orientalism, and Identity: The Politics of the Israeli-Built Environment' *Israel Studies* 13(1): 94–118.

Yacobi, Haim (2009). *The Jewish-Arab City: Spatio-Politics in a Mixed Community* (London: Routledge).

Yaacov, Ro'i (1981). 'Jewish-Arab Relations in the First Aliyah Settlements' in Mordechai Eliav (ed.) *The First Aliyah: Volume One* (Yad Ben-Zvi: Ministry of Defence Jerusalem).

Younes, Adva Him and Shira Malka (2006). *Developing a Curriculum in Arabic for Intermediate and High Schools in the Jewish Sector* (Jerusalem: Henrietta Szold Institute – The National Institute for Research in the Behavioural Sciences and the Pedagogic Secretariat in the Ministry of Education) [in Hebrew].

Zerubavel, Yael (1991). 'The Politics of Interpretation: Tel Hai in Israel's Collective Memory' *Association of Jewish Studies Review* 16(1/2): 130–60.

Zerubavel, Yael (1997). *Recovered Roots: Collective Memory and the Making of Israeli National Tradition* (Chicago: University of Chicago Press).

Zerubavel, Yael (2008). 'Memory, the Rebirth of the Native, and the "Hebrew Bedouin" Identity' *Social Research* 75(1): 315–52.

Zimenavoda, Tamar (1981). *From Grandmother's Kitchen: Erets Yisrael Dishes – New and Old* (Tel Aviv, Milo Publishers) [in Hebrew].

Zohar, Zion (2005). *Sephardic and Mizrahi Jewry* (New York: New York University Press).

Zuckermann, Ghil'ad (2008). *Israelit Safa Yafa* (*Israeli – A Beautiful Language*) (Tel Aviv: Am Oved) [in Hebrew].

Newspaper articles

Aderet, Ofer [in Hebrew] 'Berlin is Far Away from Being Paradise for Israelis', *Haaretz*, 15 October 2014: http://www.haaretz.co.il/news/education/.premium-1.2459649 (accessed: 30 March 2015).

Al-Rajoub, Oud [in Arabic] 'The Israeli Occupation Chooses *Shaqāiq al-Nuʿmān* and *Qarn al-Ghazāl* to Represent it Abroad', *Aljazeera*, 5 October 2007: http://www.aljazeera.net/home/print/f6451603–4dff-4ca1–9c10–122741d17432/e724a200-aec9–4278-b413–2fe7e83bdfbc (accessed: 18 October 2014).

Alperin, Michele 'Scholar explores Hebrew's Debt to Arabic', *New Jersey Jewish News*, 8.12.2.2014: http://njjewishnews.com/article/25502/scholar-explores-hebrews-debt-to-arabic#.VQ2PqOF8vdE (accessed: 21 March 2014).

Amit, Tal [in Hebrew] 'What is Israeli for You: The Graduates of 2014', *Design Museum Holon*, 19 June 2014: http://www.dmh.org.il/heb/magazine/magazine.aspx?id=224&IssuesId=13.

Ansky, Sherri [in Hebrew] 'The Mother of the Wheat: Introducing the Freekeh, the Mediterranean answer to Quinoa', *NRG*, 28 February 2009: http://www.nrg. co.il/online/55/ART1/858/492.html (accessed: 27 April 2015).

Arad, Boaz [in Hebrew] 'This is the Person Behind the Milkey Protest', *Haaretz*, 15 October 2014: http://www.haaretz.co.il/news/education/.premium-1.2459652 (accessed: 30 March 2015).

Atwan, Shahar [in Hebrew] 'A Holiday in the Display Window', *Haaretz*, 20 June 2012 http://www.haaretz.co.il/gallery/fashion/collection-review/1.1735582 (accessed 19 February 2015).

Barzilai, Einar and Yoel Glazer [in Hebrew] 'What is Israeli for You?: The Channel 2 viewers choose', *Mako*, 16 April 2013: http://www.mako.co.il/news-israel/ local/Article-6dd5bdee7e31e31004.htm (accessed: 18 October 2014).

Bat-Yaar, Nurit [in Hebrew] *Fashion Art*: http://nuritbatyaar-fashionart.blogspot. co.il/2010/08/israeli-fashion-kafiya-andre-courrege.html (accessed: 2 January 2015).

BBC (2008). Cooking in the Danger Zone: Israel and Palestinian Territories. Available at: http://news.bbc.co.uk/1/shared/bsp/hi/pdfs/30_03_08_cooking_in_ the_danger_zone_srs_3_israel_palestine_territories.pdf (accessed 27 February 2013).

Blum, Amalya [in Hebrew] 'You Can say Arab', *Haaretz*, 29 March 2012: http:// www.haaretz.co.il/opinions/1.1674645 (accessed: 31 January 2015).

Dafni, Amots and Salih Aql Khatib [in Hebrew] 'On the Arabic Names of the Raḵefet', *The Wildflowers of Israel Website*: http://www.wildflowers.co.il/ hebrew/tiulimReadMore.asp?ID=481 (accessed: 27 April 2015).

Dror, Yuval 'Without Borders', *Haaretz*, 14 September 2004: http://www.haaretz. com/without-borders-1.134751 (accessed 31 January 2015).

Dror, Yuval [in Hebrew] 'There is Nothing like Jaffa in the World', *Haaretz*, 12 September 2004: http://www.haaretz.co.il/news/health/1.999664 (accessed 31 January 2015).

Gerty, Yael [in Hebrew] 'Freekeh: Smoked, Green and Cool', *Ynet*, 10 December 2007: http://www.ynet.co.il/articles/0,7340,L-3480578,00.html (accessed: 27 April 2015).

Golan, Tiki [in Hebrew] 'What is the Most Israeli Dish?', *'Akhbar Ha-'Ir*, 5 May 2008.

Go Jerusalem, The Top Five Breakfast Options in Jerusalem. Available at: www. gojerusalem.com/discover/article_1476/The-top-five-breakfast-options-in-Jerusalem (accessed 27 February 2013).

Guttman, Vered 'Introducing Freekeh, the Hip Grain of 2012', *Haaretz*, 19 January 2012: http://www.haaretz.com/blogs/modern-manna/introducing-freekeh-the-hip-grain-of-2012-1.408155 (accessed: 27 April 2015).

Halutz, Doron [in Hebrew] 'Haya Molcho Ccarries the Message to the Europeans: Hummus, Zionism and Determination', *Haaretz*, 18 September 2014: http:// www.haaretz.co.il/magazine/.premium-1.2437548 (accessed: 26 March 2015).

Hasan, Roee [in Hebrew] 'I look like an Arab', *Ynet*, 14 November 2014: http://www. mynet.co.il/articles/0,7340,L-4453409,00.html (accessed: 26 March 2015).

HowtobeIsraeli, (2009). 'How to Make Israeli Salad'. Available at: http://howtobeisraeli. blogspot.ca/2009/02/lesson-2-make-israeli-salad.html (accessed 10 June 2014).

Katzir, Ronnie [in Hebrew] 'Fashion: Homage to Israeli Icons', *Calcalist*, 2 September 2013: http://www.calcalist.co.il/consumer/articles/0,7340,L-3611546,00.html (accessed 21 February 2015).

Klein, Yossi [in Hebrew] 'Look! An Arab speaking Hebrew, *Haaretz*, 12 July 2013.

Klein, Zeev 'Ayalon: Israeli, Palestinian Economies – Inseparable Siamese Twins', *Globes*, 4 December 2000: http://www.globes.co.il/en/article-454783 (accessed: 2 April 2015).

Lee, Vered [in Hebrew] 'Five Questions to Dan Ronen', *Haaretz*, 5 April 2011: http://www.haaretz.co.il/literature/1.1173013 (accessed: 26 March 2015).

Lior, Ilan [in Hebrew] 'Who will Watch the Canaanite Watchdogs?', *Haaretz* 6 April 2012: http://www.haaretz.co.il/news/science/1.1680938 (accessed: 29 September 2015).

Madar, Revital [in Hebrew] 'Mizrahi, Ashkenazi, Palestinian Wearing Galabiyya: Who has the Cultural Right to Wear It?', *Haaretz*, 5 June 2015: http://www. haaretz.co.il/gallery/black-flag/.premium-1.2651120 (accessed: 6 June 2015).

McGrane, Sally 'So Long Israel, Hello Berlin', *The New Yorker*, 15 May 2014: http://www.newyorker.com/culture/culture-desk/so-long-israel-hello-berlin (accessed: 30 March 2015).

Murad, Eyal [in Hebrew] 'He is One of Us: On the Israeli Cnaan Dog', *NRG*, 2 May 2006: http://www.nrg.co.il/online/1/ART1/170/115.html (accessed: 27 April 2015).

Raved, Ahiya [in Hebrew] 'Youth Believe Arabs Dirty', *Ynetnews*, 1 September 2007: http://www.ynet.co.il/articles/0,7340,L-3350424,00.html (accessed: 23 April 2015).

'A Rental Car Blue and White' [in Hebrew] *Maariv*, 14 January 1987: http:// jpress.org.il/Olive/APA/NLI_heb/?action=tab&tab=browse&pub=MAR (accessed: 18 October 2014).

Sales, Ben 'After Gaza conflict, Israel's Arabs Fear Rising Discrimination', *Haaretz*, 7 September 2014.

Shaked, Sharon [in Hebrew] 'For Our Independence: An Israeli Food Dictionary', *Fresh*, 5 May 2003.

Shalev, Meir [in Hebrew] 'The Hummus is Ours', *Yedioth Aharonoth*, 12 January 2001.

Stern, Yoav [in Hebrew] 'Poll: 50% of Israeli Jews Support State-Backed Arab Emigration', *Haaretz*, 27 March 2007.

'The "Blue and White" Campaign of Eldan on Shirts Made in Turkey', *TheMarker*, 14 April 2010: http://www.themarker.com/misc/1.575877 (accessed: 18 October 2014).

The Official Website of *Wild Flower of Israel* [in Hebrew]. Available at: http://www.wildflowers.co.il/hebrew/plant.asp?ID=61 (accessed: 1 February 2015).

'The Racial Slur Database', Available at: http://www.rsdb.org/race/arabs (accessed: 31 January 2015).

Yaakov, Itay [in Hebrew] 'The Fashion Academy of Yael Guilat', *Ynet*, 3 January 2012: http://xnet.ynet.co.il/fashion/articles/0,14539,L-3092334,00.html (accessed 20 June 2015).

Yaron, Oded [in Hebrew] 'Israeli Hip-Hop Star Hopping Mad over 'Palestine' Gazelle', *Haaretz*, 28 April 2015: http://www.haaretz.com/news/national/.premium-1.654011 (accessed 28 April 2015).

Zaka'im, Har'el [in Hebrew] 'Preparing for Independence: The Guide for the Perfect Musabaha', *Haaretz* 19 April 2015: http://www.haaretz.co.il/gallery/recipes/mangal/1.2617312/ (accessed: 23 April 2015)

Index

Abbas, Hussam 77–8, 87, 95
Abu Shaqra, ʿĀṣim 57
Acre 40, 112, 131
Afula 85
Al-Babūr (restaurant) 77–8, 87, 96
Albukhārī 40
ʾAliya (Jewish immigration to Palestine/
    Israel) xiii, xxi, 9, 38, 66, 75, 90
Al-Saʿdi, Zulfa 56
Al-Shahristānī 40
Al-Ṭabarīi 40
Alterman, Nathan 64
Altneuland ix, 8, 30
Americanisation 103, 105, 124
Andromeda Hill 131–2
Anti-Semitism 7, 76
Arabic (language) xiv, xix, 4, 11–12,
    15–18, 20–24, 26–32, 35–46, 54,
    56, 62–5, 69, 88, 91–2, 119, 122,
    127, 137
Arabic studies 16, 26–9, 35–41, 44–6, 91
Arab-Israeli conflict xv, 2, 4, 7, 28–30, 36,
    38, 43, 58, 60, 65, 70, 76, 79–80,
    91–2, 101, 112, 114, 121, 135
Arab-Jew (see also: Mizrahi-Jew) xii, 10,
    23–4, 28, 124, 126, 138, 141
Arab, Image of the xi–xii, 46, 109–15
Arab-Islamic empires 21
Arab-Palestinians
    citizens of Israel xi, 3–4, 45, 57–8,
        73–4, 77–9, 87–8, 96–101, 112–15,
        117, 119, 130–32, 135–6, 138–41
    folk traditions 62–4, 110, 117–19, 122
    food x, 1, 3–4, 12–13, 77–9, 85–101,
        139
    herbs 91, 94, 97–8, 122
    music 117, 119, 122
    nationalism 47–8, 56–9, 72, 98, 113
    orientalisation of xi–xiii, 7–8, 10–11,
        29–30, 87, 111–13, 117, 124–5

restaurants 3, 77–9, 87, 96
urban population 3, 30, 90, 112, 136
villages 36, 48, 56, 60, 79–80, 87,
    130–31
Arab Revolt 39, 41
Arab world x, xii, 13, 17, 39, 45, 86, 94,
    101, 124, 138
Aramaic (language) 20, 42–3, 61–2, 92
Architecture 13, 127–32
Ariel, Meir, xii
Ashkenazi community in historical
    Palestine 25–9, 89–90
Ashkenazi Jews (also European Jewry) xi,
    5, 7–8, 10, 29, 44, 105, 108–9, 114,
    135, 138
Ashkenazi Hebrew pronunciation 41–4

Baerwald, Alexander 127–8
Baghdad 20–21
Banal nationalism 5, 51, 81, 94, 108
Barak, Ehud xi, 93
Barakeh, Muhammad 77, 80
Bar-Giyora (organisation) 31
Bat Yam 104–5
Bedouin xiii, xxi, 11, 30–32, 36, 38, 60,
    68–72, 74–5, 90–91, 96, 101,
    111–12, 122–4, 136
Beijing 61, 64, 141
Ben-Gurion, David ix, 30, 67, 70, 72–4,
    91, 128
Ben-Yehouda, Eliʿezer 12, 35–6, 40–43, 46
Ben-Yehouda, Netiva 121
Bentov, Mordechai 129
Berlin 103–5
Bhabha, Homi 4, 7
Bible, the Old Testament (also Jewish
    biblical heritage) 9–11, 13, 20, 23,
    30, 36, 43–4, 47, 51–2, 54, 59–61,
    65–8, 70, 75, 83, 87, 90, 111,
    125–6, 129–30, 132–3, 136–7